MEDIA AND COMMUNICATIONS - TECHNOLOGIES, POLICIES AND CHALLENGES

MOBILE DEVICES

PRIVACY RISKS AND PROTECTIONS

MEDIA AND COMMUNICATIONS - TECHNOLOGIES, POLICIES AND CHALLENGES

Additional books in this series can be found on Nova's website under the Series tab.

Additional E-books in this series can be found on Nova's website under the E-book tab.

MEDIA AND COMMUNICATIONS - TECHNOLOGIES, POLICIES AND CHALLENGES

MOBILE DEVICES

PRIVACY RISKS AND PROTECTIONS

ALLEN R. WALKER
AND
DAVIS BROWN MILLER
EDITORS

nova publishers
New York

For permission to use material from this book please contact us:
Telephone 631-231-7269; Fax 631-231-8175
Web Site: http://www.novapublishers.com

Library of Congress Cataloging-in-Publication Data

ISBN: 978-1-62618-304-9

Published by Nova Science Publishers, Inc. † New York

CONTENTS

PREFACE

Smartphones can provide services based on consumers' location, raising potential privacy risks if companies use or share location data without consumers' knowledge. The Federal Trade Commission (FTC) enforces prohibitions against unfair and deceptive practices, and the National Telecommunications and Information Administration (NTIA) sets national telecommunications policy. This book addresses how mobile industry companies collect location data, why they use and share these data, and how this affects consumers. Also discussed are the types of actions private sector entities have taken to protect consumers' privacy and ensure security of location data.

Chapter 1 – Smartphones can provide services based on consumers' location, raising potential privacy risks if companies use or share location data without consumers' knowledge. FTC enforces prohibitions against unfair and deceptive practices, and NTIA sets national telecommunications policy. GAO was asked to examine this issue. GAO reviewed (1) how mobile industry companies collect location data, why they share these data, and how this affects consumers; (2) actions private sector entities have taken to protect consumers' privacy and ensure security of location data; and (3) actions federal agencies have taken to protect consumer privacy and what additional federal efforts, if any, are needed. GAO analyzed policies and interviewed representatives of mobile industry companies, reviewed documents and interviewed officials from federal agencies, and interviewed representatives from industry associations and privacy advocates.

Chapter 2 – This is the Statement of Jessica Rich, Deputy Director, Bureau of Consumer Protection, Federal Trade Commission. Hearing on "Protecting Mobile Privacy: Your Smartphones, Tablets, Cell Phones and Your Privacy."

Chapter 3 – This is the Statement of Justin Brookman, Director, Consumer Privacy, Center for Democracy and Technology. Hearing on "Protecting Mobile Privacy: Your Smartphones, Tablets, Cell Phones and Your Privacy."

Chapter 4 – This is the Testimony of Ashkan Soltani, Independent Researcher and Consultant. Hearing on "Protecting Mobile Privacy: Your Smartphones, Tablets, Cell Phones and Your Privacy."

Chapter 5 – This is the Testimony of Alan Davidson, Director of Public Policy, Google. Hearing on "Protecting Mobile Privacy: Your Smartphones, Tablets, Cell Phones and Your Privacy."

Chapter 6 – This is the Testimony of Dr. Guy "Bud" Tribble, Vice President for Software Technology, Apple Inc.. Hearing on "Protecting Mobile Privacy: Your Smartphones, Tablets, Cell Phones and Your Privacy."

Chapter 7 – This is the Statement of Jonathan Zuck, President, The Association for Competitive Technology. Hearing on "Protecting Mobile Privacy: Your Smartphones, Tablets, Cell Phones and Your Privacy."

In: Mobile Devices ISBN: 978-1-62618-304-9
Editors: A. R. Walker, D. Brown Miller © 2013 Nova Science Publishers, Inc.

Chapter 1

MOBILE DEVICE LOCATION DATA: ADDITIONAL FEDERAL ACTIONS COULD HELP PROTECT CONSUMER PRIVACY*

United States Government Accountability Office

WHY GAO DID THIS STUDY

Smartphones can provide services based on consumers' location, raising potential privacy risks if companies use or share location data without consumers' knowledge. FTC enforces prohibitions against unfair and deceptive practices, and NTIA sets national telecommunications policy. GAO was asked to examine this issue. GAO reviewed (1) how mobile industry companies collect location data, why they share these data, and how this affects consumers; (2) actions private sector entities have taken to protect consumers' privacy and ensure security of location data; and (3) actions federal agencies have taken to protect consumer privacy and what additional federal efforts, if any, are needed. GAO analyzed policies and interviewed representatives of mobile industry companies, reviewed documents and interviewed officials from federal agencies, and interviewed representatives from industry associations and privacy advocates.

---

* This is an edited, reformatted and augmented version of United States Government Accountability Office, Publication No. GAO-12-903, dated September 2012.

WHAT GAO RECOMMENDS

GAO recommends that NTIA work with stakeholders to outline specific goals, milestones, and performance measures for its process to develop industry codes of conduct and that FTC consider issuing guidance on mobile companies' appropriate actions to protect location data privacy. Because the agencies had concerns about certain aspects of GAO's draft recommendations, GAO revised them by including that NTIA should work with stakeholders in the process to develop industry codes and removing from the draft FTC recommendation that the guidance should include how FTC will enforce the prohibition against unfair practices.

WHAT GAO FOUND

Using several methods of varying precision, mobile industry companies collect location data and use or share that data to provide users with location-based services, offer improved services, and increase revenue through targeted advertising. Location-based services provide consumers access to applications such as real-time navigation aids, access to free or reduced-cost mobile applications, and faster response from emergency services, among other potential benefits. However, the collection and sharing of location data also pose privacy risks. Specifically, privacy advocates said that consumers: (1) are generally unaware of how their location data are shared with and used by third parties; (2) could be subject to increased surveillance when location data are shared with law enforcement; and (3) could be at higher risk of identity theft or threats to personal safety when companies retain location data for long periods or share data with third parties that do not adequately protect them.

Industry associations and privacy advocates have developed recommended practices for companies to protect consumers' privacy while using mobile location data, but companies have not consistently implemented such practices. Recommended practices include clearly disclosing to consumers that a company is collecting location data and how it will use them, as well as identifying third parties that companies share location data with and the reasons for doing so. Companies GAO examined disclosed in their privacy policies that the companies were collecting consumers' location data, but did not clearly state how the companies were using these data or what third parties they may share them with. For example, some companies' policies stated they collected location data and listed uses for personal information, but did not state clearly whether companies considered location to be personal information. Furthermore, although policies stated that companies shared location data with third parties, they were sometimes vague about which types of companies these were and why they were sharing the data. Lacking clear information, consumers faced with making a decision about whether to allow companies to collect, use, and share data on their location would be unable to effectively judge whether the uses of their location data might violate their privacy.

Federal agencies have held educational outreach events, developed reports with recommendations aimed at protecting consumer privacy, and developed some guidance on certain aspects of mobile privacy. The Department of Commerce's National Telecommunications and Information Administration (NTIA) is implementing an administration-proposed effort to bring industry, advocacy, and government stakeholders together to develop codes of conduct for industry to address Internet consumer privacy issues generally. However, NTIA has not set specific goals, milestones, and performance measures for this effort. Consequently, it is unclear if or when the process would address mobile location privacy. Furthermore, the Federal Trade Commission (FTC) could enforce adherence to the codes if companies adopted them, but since adoption is voluntary, there is no guarantee companies would adopt the resulting codes. While FTC has issued some guidance to address mobile location privacy issues, it has not issued comprehensive guidance that could inform companies of the Commission's views on the appropriate actions companies should take to protect consumers' mobile location data privacy.

ABBREVIATIONS

ACLU	American Civil Liberties Union
A-GPS	Assisted Global Positioning System
AMBER	America's Missing: Broadcast Emergency Response
Commerce	Department of Commerce
Communications Act	Communications Act of 1934
CPNI	customer proprietary network information
CTIA	CTIA-The Wireless Association
ECPA	Electronic Communications Privacy Act of 1986

E911	enhanced 911
FCC	Federal Communications Commission
FIP	Fair Information Practice
FTC	Federal Trade Commission
GPS	Global Positioning System
Justice	Department of Justice
NTIA	National Telecommunications and Information Administration
OECD	Organisation for Economic Co-operation and Development
Wi-Fi	wireless fidelity

September 11, 2012

The Honorable Al Franken
Chairman
Subcommittee on Privacy, Technology and the Law
Committee on the Judiciary
United States Senate

Dear Mr. Chairman:

The number of mobile phone subscriptions in the United States grew from about 3.5 million in 1989 to approximately 291 million by the end of 2009, according to the most recent Federal Communications Commission (FCC) data.[1] The Centers for Disease Control and Prevention reported that one-third of U.S. households had mobile phones but no landline phones as of December 2011.[2] Increasingly, Americans' mobile phones are smartphones, which use advanced operating systems to provide computing functions, including Internet access and a variety of applications, in addition to basic voice service. According to The Nielsen Company, as of June 2012, smartphones accounted for just over half of all mobile phones in the United States, up from less than one-quarter in early 2010.[3]

Smartphones allow users access to location-based services that can provide them with navigation tools and information relevant to their surroundings based on increasingly precise information about the user's current location determined by Global Positioning System (GPS) and other methods. Location-based services have proved popular with users; the Pew Research Center reported that three-quarters of smartphone users were using such services as of February 2012.[4] In providing such services, smartphones and the companies that support their functions are able to collect and retain precise data about users' locations. Concerns have been raised about how mobile industry companies that provide or enable location-based services use and share consumers' location data, raising the potential that consumers' privacy could be violated if their location data are used in ways they did not intend or authorize.

Several agencies have responsibility to address mobile phone consumers' privacy and create related guidance. The Federal Trade Commission (FTC) has authority to take enforcement actions against unfair or deceptive acts or practices of companies; FCC has regulatory and enforcement authority over mobile carriers, such as AT&T and Verizon; and the Department of Commerce's (Commerce) National Telecommunications and Information Administration (NTIA) advises the President on telecommunications and information policy issues. Additionally, the Department of Justice (Justice) disseminates guidance on procedures for law enforcement to request electronic evidence, such as a person's current or historical location data.

This report addresses

1. how mobile industry companies collect location data, why they use and share these data, and how this affects consumers;
2. the types of actions private sector entities have taken to protect consumers' privacy and ensure security of location data; and
3. the actions federal agencies have taken to protect consumer privacy and what additional federal efforts, if any, are needed.

To address these issues, we examined privacy policies and interviewed representatives from carriers, operating system developers, and smartphone manufacturers that are the largest in the United States by market share and representatives from the developers of the most popular mobile applications on the Apple and Google operating systems.[5] We reviewed documents and interviewed officials from FTC, FCC, NTIA, and Justice. We also interviewed privacy advocates and industry association representatives on their views. In considering ways to address location data privacy issues, we are reporting actions federal agencies could take, rather than potential legislative options. Recent proposals in Congress have sought to amend how the law treats mobile device location data, which is among the technology issues that have recently emerged.[6] See appendix I for a more detailed description of our scope and methodology.[7]

We conducted this performance audit from December 2011 to September 2012 in accordance with generally accepted government auditing standards. Those standards require that we plan and perform the audit to obtain sufficient, appropriate evidence to provide a reasonable basis for our findings and conclusions based on our audit objectives. We believe that the evidence obtained provides a reasonable basis for our findings and conclusions based on our audit objectives.

BACKGROUND

Smartphones combine the telecommunications functions of a mobile phone with the processing power of a computer, creating an Internet-connected mobile device capable of running a variety of software applications for productivity or leisure. The functioning of a mobile phone involves locating the user,[8] and FCC's rules enabling enhanced 911 (E911) services require phones to provide GPS-quality location precision for emergency responders. This capability to determine a user's location has led to smartphones that can provide applications and services that take advantage of location data generated by GPS and other location technologies. Advances in the technology for pinpointing a mobile phone's location have led to applications that identify a user's location quickly and with a high-level of precision.

Four types of companies are primarily responsible for smartphone products and services in the United States:

- Mobile carriers. Carriers provide smartphone users with access to wireless networks for voice and data uses, generally with a subscription plan. In the United States, four carriers primarily serve customers nationwide: AT&T, Sprint-Nextel, T-Mobile, and Verizon.
- Operating systems. Underlying the various functions of a smartphone is an operating system that acts as a mobile computing platform to run the phone's hardware and software. Three operating systems are most prevalent in the United States: Apple's iPhone iOS, Google's Android, and Research in Motion's BlackBerry.
- Manufacturers. Smartphones are made by a variety of electronics companies. Apple and Research in Motion manufacture phones based on their own proprietary operating systems. In contrast, a number of

other companies, such as HTC, Motorola, and Samsung, make phones based on the Android operating system.

- Application developers. As the popularity of smartphones has grown, so too has the number of developers offering applications for them. New mobile applications are developed every day, with some estimates indicating there are more than a million available as of mid-2012. These developers range from start-up ventures to large, established Internet companies like Yahoo!, offering products like the Angry Birds game by Rovio Entertainment Ltd., social networking applications like Facebook, navigation tools like Google Maps, and music players such as Pandora Radio.

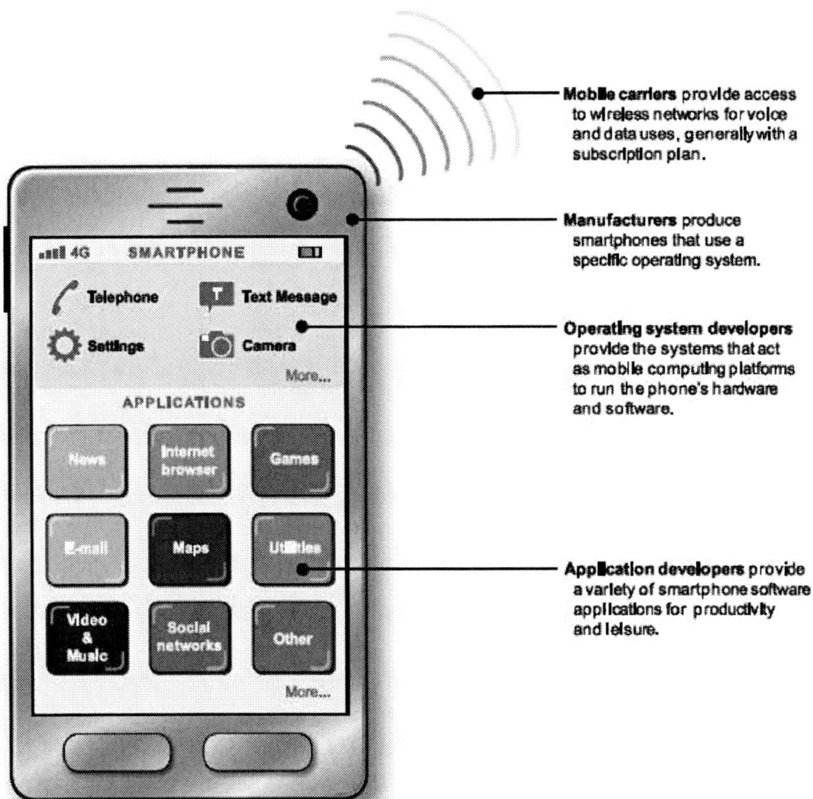

Source: GAO.

Figure 1. Roles of Mobile Industry Companies in Delivering Smartphone Functions.

Together, the products and services developed by these various companies allow users to take advantage of the various functions smartphones provide (see figure 1).

Smartphones connect with mobile carrier networks for making calls and providing data services. Some smartphones also have the capability to connect with wireless fidelity (Wi-Fi) networks to provide data services.

Fair Information Practices

Fair Information Practices (FIP), are widely accepted principles for protecting the privacy and security of personal information. They were first proposed in 1973 by a U.S. government advisory committee. In response to concerns about the potential consequences that computerized data systems could have on the privacy of personal information, the committee was tasked to examine the extent to which limitations should be placed on using computer technology for record keeping about people.[9] These principles, with some variation, have been used by organizations to address privacy considerations in their business practices and are also the basis of privacy laws and related policies in many countries, including the United States. FIPs are not precise legal requirements. Rather, they provide a framework of principles for balancing the need for privacy with other interests. Striking that balance varies among countries and among types of information (e.g., medical and employment information).

The Organisation for Economic Co-operation and Development (OECD), an international organization, developed a revised version of the FIPs in 1980 that has been widely adopted (see table 1).[10]

Laws that Govern Private-Sector Use of Personal Information

The Federal Trade Commission Act prohibits unfair or deceptive acts or practices affecting commerce and authorizes FTC enforcement action.[11] This authority allows FTC to take remedial action against a company that engages in a practice that FTC has found is unfair or deceives customers. For example, FTC could take action against a company if it found the company was not adhering to the practices to protect a consumer's personal information that the company claimed to abide by in its privacy policy. FTC also enforces the Children's Online Privacy Protection Act of 1998, which required FTC to promulgate rules governing the online collection of information from children under age 13.[12]

Table 1. The OECD Fair Information Practices

Principle	Description
Collection limitation	The collection of personal information should be limited, should be obtained by lawful and fair means, and, where appropriate, with the knowledge or consent of the individual.
Data quality	Personal information should be relevant to the purpose for which it is collected, and should be accurate,complete, and current as needed for that purpose.
Purpose specification	The purposes for the collection of personal information should be disclosed before collection and upon any change to those purposes, and the use of the information should be limited to those purposes and compatible purposes.
Use limitation	Personal information should not be disclosed or otherwise used for other than a specified purpose without consent of the individual or legal authority.
Security safeguards	Personal information should be protected with reasonable security safeguards against risks such as loss or unauthorized access, destruction, use, modification, or disclosure.
Openness	The public should be informed about privacy policies and practices, and individuals should have ready means of learning about the use of personal information.
Individual participation	Individuals should have the following rights: to know about the collection of personal information, to access that information, to request correction, and to challenge the denial of those rights.
Accountability	Individuals controlling the collection or use of personal information should be accountable for taking steps to ensure the implementation of these principles.

Source: OECD.

The Communications Act of 1934 (Communications Act), as amended, imposes a duty on mobile carriers to secure information and imposes particular requirements for protecting information identified as customer proprietary network information (CPNI), including the location of customers when they make calls.[13] The Communications Act requires express authorization for access to or disclosure of call location information concerning the user of commercial mobile services, subject to certain exceptions.[14] Carriers must also comply with FCC rules implementing the E911 requirements of the Wireless Communications and Public Safety Act of 1999,[15] including providing location information to emergency responders when mobile phone users dial 911.[16]

The Electronic Communications Privacy Act of 1986 (ECPA) sets out requirements under which the government can access information about a user's mobile phone and Internet communications.[17] This includes legal procedures for obtaining court orders to acquire information relevant to a law enforcement inquiry.

Companies Collect, Use, and Share Location Data That Provide Consumer Benefits, but Also Pose Privacy Risks

Collecting, using, and sharing location data provides benefits for both mobile industry companies and for consumers. For the companies, the main purposes for using and sharing location data are to provide and improve services, to increase advertising revenue, and to comply with legal requirements. Consumers, in turn, can benefit from these new and improved services and from targeted location-based advertising. Nonetheless, allowing companies to access location data exposes consumers to privacy risks, including disclosing data to unknown third parties for unspecified uses, consumer tracking, identity theft, threats to personal safety, and surveillance.

Companies Collect Location Data in Various Ways

Mobile industry companies determine location information through various methods, such as cell tower signal-based technologies, Wi-Fi Internet access point technology, crowd-sourced positioning, and GPS technology. Assisted-GPS (A-GPS), a hybrid technology that uses more than one data collection methodology, is also widely used. Figure 2 below illustrates these technologies.

Cell Tower Signal-Based Technologies

Since the advent of consumer cellular technology, making and receiving mobile telephone calls has depended on the ability to determine a device's location from the constant radio communication between the device and the mobile carrier's cell towers that are spread throughout the carrier's service area. The ranges of the individual cell towers divide the service area into separate sectors. As the towers are in fixed positions, determining a device's current cell tower sector tells the carrier the device's approximate location. The precision of this method depends on how much space a particular tower covers.

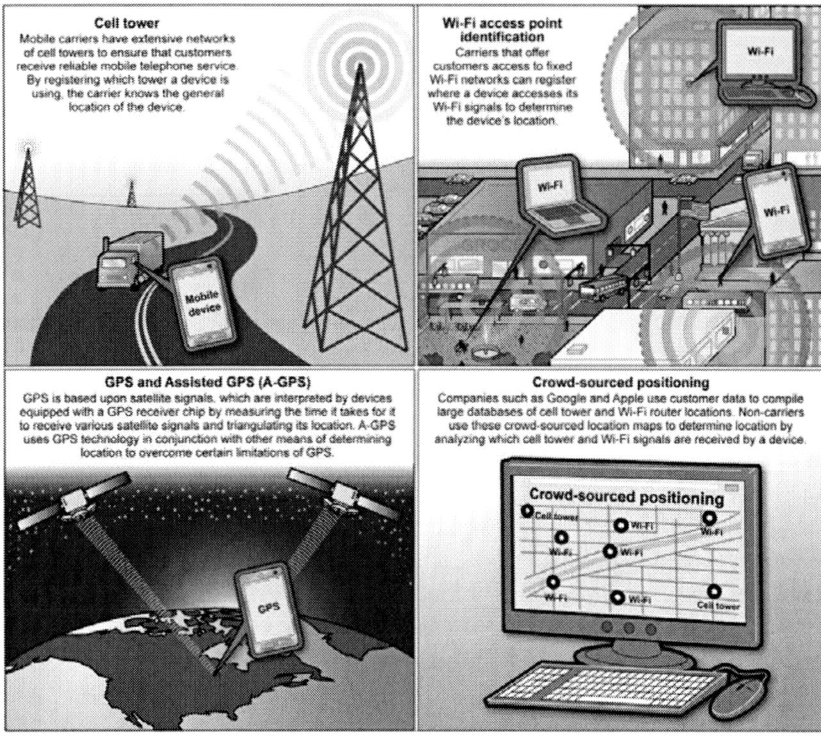

Source: GAO.

Figure 2. Methods Used to Collect Location Information.

In general, urban areas have smaller sectors than rural areas because each sector can only manage a certain amount of cell traffic at any one time. Because of increasing cell traffic, the number of cell towers has proliferated to the point that there are now over three times more than there were 10 years ago. As a result, cell sector-based location data are increasingly accurate. Companies can further improve accuracy by using triangulation methods, which determine location through the mathematical comparison of a device's signals that reach more than one cell tower. Cell tower triangulation can now yield results within 50 meters of accuracy.

Wi-Fi Access Point Identification

Mobile carriers that provide Wi-Fi access points to their customers can use these access points to determine location. Like cell towers, Wi-Fi access points are fixed locations and send out signals over a limited range. Specifically, Wi-Fi signals are radio waves that provide Internet access to devices equipped with compatible wireless hardware. Each WiFi access point is identified by a unique hardware address. Nearby compatible devices are able to receive this information and use it to request Internet access. Since a Wi-Fi access point's range is limited to a few hundred meters, accurate location data can be determined if a device communicates with the access point.

Crowd-Sourced Positioning

Companies such as Google, Apple, and Skyhook use information gathered from users' mobile devices about cell tower and Wi-Fi access point signals, as well as the Wi-Fi signals of other companies and households, to determine location. These companies compile the precise locations of these signals into large databases, which the companies may then license to other entities such as application developers. An application installed on a mobile device can obtain location information by querying one of these databases, which will use its knowledge about those signals' locations to return the device's location. The database can also use location information sent by the device to update its records. If there are any new signals in the device's vicinity or any old signals that are no longer broadcasting, the database can incorporate those changes in its records. While the exact degree of accuracy ultimately depends on how many signal points are near the device when it queries a database, companies use crowd-sourced positioning because it provides accurate location data quickly, and because it does not rely on GPS technology, which is not available in all mobile devices.

GPS and A-GPS

GPS is used by both carriers and non-carriers to determine a device's location. GPS technology is based upon satellite signals, which are picked up and interpreted by devices equipped with GPS receiver chips. The device then measures the time it takes for it to receive various satellite signals and triangulates its location. Triangulating GPS satellite signals can yield data accurate to within 10 meters.

A-GPS is a hybrid approach used to overcome certain limitations in GPS technology: namely, that GPS usually only works outside buildings, may take several minutes to determine location, and uses more battery power than other location determination methods. By using GPS in conjunction with any of the previously described methods of collecting location data, the assisting technology can report an approximate location to the application or service while GPS works to obtain a more precise location. For instance, operating system and application developers may use crowd-sourced positioning databases to provide approximate locations to their users until GPS signals are successfully triangulated. The precision of A-GPS in these circumstances depends on the accuracy of the assisting method.

Mobile Industry Companies Use and Share Location Data for Various Reasons

There are three main reasons that mobile industry companies collect and share location data: 1) to provide and improve services, 2) to increase advertising revenue, and 3) to comply with court orders.

Provide and Improve Services

Mobile industry companies use location data to provide and improve services. As stated above, a carrier needs to know a device's location to provide basic mobile telephone services. In addition, carriers and application developers offer a diverse array of services that make use of location information, such as services providing navigation, the ability to keep track of family members, local weather forecasts, the ability to identify and locate nearby businesses, and social networking services that are linked to users' locations. To provide these services, carriers and developers need the ability to quickly and accurately determine location. Location data can also be used to enhance the functionality of other services that do not need to know the user's location to operate. Search engines, for example, can use location data as a frame of reference to return results that might be more relevant. For instance, if a user were to search for a pizza restaurant using a location-aware search engine, the top result may be a map of nearby pizza restaurants instead of the homepage of a national chain.

Companies also collect and examine location information in conjunction with other diagnostic usage data to analyze and improve their interactions with customers. By examining the location patterns of dropped calls, for example, carriers can identify network problems and address cell connectivity issues without having to rely on customer complaints.

Furthermore, companies may use location data to provide public services. For example, carriers are responsible for providing law enforcement and other first responders with the location data of people who dial 911 from their mobile devices. This service is referred to as E911 and it is mandated by law.[18] In addition, companies may provide location information to municipalities to improve city traffic management or facilitate city planning. Location data can also be used to help find missing children through mobile America's Missing: Broadcast Emergency Response (AMBER) alerts,[19] which can be sent to devices that have requested AMBER alerts, when the devices are located within a specified radius of a reported incident.

Increase Advertising Revenue

Companies can use location data to target the advertising that users receive through mobile devices. Doing so may make an advertisement more relevant to a user than a non-targeted advertisement, boosting advertising revenue. Advertising is particularly important to application developers, as many developers give their products away free and rely on advertising for revenue. Advertisements for a certain business may be triggered if a user's device is located within a predetermined distance from that business. Any application, regardless of its function, may collect and use location data for advertising purposes.

Furthermore, application developers, operating system developers, and mobile carriers may aggregate and store individual user data to create user profiles. Profiles can be used to tailor marketing or service performance to an individual's preferences. In addition to capturing and using the location data of individual users, companies such as application developers and mobile carriers sell large amounts of de-identified location data to third parties. When data are de-identified, they are stripped of personally identifiable information.[20] In addition to de-identification, user data are often aggregated, which means that the data of many users are combined. Aggregation also makes it more difficult to distinguish the data of individuals. De-identified and aggregated data can be used for a variety of purposes, including marketing and research.

Comply with Court Orders

Mobile industry companies are legally required to share user location data in response to a court order if a court finds that the information is warranted for law enforcement purposes. Because users generally carry their mobile devices with them, law enforcement can use device location data to determine the user's location. Because of this correlation, location data are valuable to law enforcement for tracking the movements of criminal suspects. Of particular use are the location data either housed in mobile carrier databases or obtained through GPS technology. Mobile carriers are required to comply with court orders directing the disclosure of historical location data (i.e., where the device was in the past) and in certain circumstances, real-time location data (i.e., where the device is now).

Location Data Use and Sharing Can Benefit Consumers, but Also Pose Privacy Risks

Many services that use location data were designed to make tasks easier or quicker for the customer, and the sharing of location data can improve customer experiences, reduce consumer costs, and help provide improved public services. Nonetheless, location data use and sharing may pose privacy risks, which include unknown third-party use, consumer tracking, identity theft, threats to personal safety, and surveillance.

Consumer Benefits

Consumers can benefit from mobile industry use of their location data because many location-based services are designed to make their lives easier and safer. For instance, navigation services enable users to easily find directions and take the guesswork out of finding the best or quickest routes, while applications designed to track family members enable parents to be aware of their children's whereabouts. An application may also use location data to personalize its usual services; for example, by using a location-aware business directory, a user may be able to rank search results by distance to save time and quickly reach the nearest location. Furthermore, as stated previously, the sharing of location data facilitates a faster response from emergency services through E911 and allows companies to identify network service problems.

Additionally, consumers may derive economic benefits from the sharing of their location data. For example, because many application developers depend on location-based advertising for revenue, users may be able to download applications for free or at a low cost. Furthermore, location-based advertising allows for targeted advertisements and offers to be sent to consumers, who may find them useful. For example, a user at lunchtime may receive and use a coupon for a local restaurant.

Consumer Privacy Risks

By allowing companies to access their location data, users expose themselves to privacy risks. These risks include, but are not limited to, disclosure to unknown third parties for unspecified uses, consumer tracking, identity theft, threats to physical safety, and surveillance.

Disclosure to Unknown Third Parties for Unspecified Uses

According to privacy advocates, when a user agrees to use a service that accesses location data, the user is unlikely to know how his or her location data may be used in ways beyond enabling the service itself. The secondary uses of location data are generally not transparent to the consumer.[21] Therefore, location data may be shared with third parties unknown to the consumer. Generally speaking, once location data are shared with a non-carrier, consumers have a limited ability to know about or influence the data's use.

Third parties that receive shared location information may vary in the levels of security protection they provide. If any of these entities has weak system protections, there is an increased likelihood that the information may be compromised. According to the congressional testimony of a privacy researcher, [22] privacy notices rarely differentiate between first- and third-party data uses and generally do not reveal specific business partners such as advertising networks, thus making it difficult for consumers to understand privacy risks. Because consumers do not know who these entities are or how they are using consumers' data, consumers may be unable to make meaningful choices and judge whether they are disclosing their data to trustworthy entities.

Tracking Consumer Behavior

When mobile location data are collected and shared, users may be tracked for marketing purposes without their consent. Since users often carry their mobile devices with them and can use them for various purposes, location data along with data collected on the device may be used to form a comprehensive record of an individual's activities. Amassing such data over time allows for the creation of a richly detailed profile of individual behavior, including habits, preferences, and routines—private information that could be exploited. Furthermore, since non-carriers' use of location data is unregulated, these companies do not have to disclose how they are using and sharing these profiles. Consumers may believe that using these personal profiles for purposes other than providing a location-based service constitutes an invasion of privacy, particularly if the use is seen as contrary to consumers' expectations and results in unwanted solicitations or other nuisances.

Identity Theft

Identity theft occurs when someone uses another person's personal or financial information to commit fraud or other crimes. When sensitive information such as location data is disclosed, particularly when it is combined with other personal information, criminals can use this information to steal identities. The risk of identity theft grows whenever entities begin to collect data profiles, especially if the information is not maintained securely. By illicitly gaining access to these profiles, criminals acquire information such as a user's name, address, interests, and friends' and co-workers' names. In addition, a combination of data elements—even elements that do not by themselves identify anyone, such as individual points of location data—could potentially be used in aggregate to discern the identity of an individual. Furthermore, keeping data long-term, particularly if it is in an identifiable profile, increases the likelihood of identity theft.

Personal Security

When mobile location data are collected and shared, users could be put at risk for personal threats if the data are intercepted by people who mean them harm. This is a potential concern for those people who do not want specific individuals to know where they are or how to find them, such as victims of domestic violence. Location data may be used to form a comprehensive record of an individual's movements and activities. If disclosed or posted, location data may be used by criminals to identify an individual's present or probable future location, particularly if the data also contain other personally identifiable information.

This knowledge may then be used to cause harm to the individual or his property through, for instance, stalking or theft. Access to location information also raises child safety concerns as more and more children access mobile devices and location-based services. According to the American Civil Liberties Union (ACLU), location updates that users provide through social media have been linked to robberies, and GPS technology has been involved in stalking cases.

Surveillance

Law enforcement agencies can obtain location data via court order, and such data can be used as evidence. However, according to a report by the ACLU, law enforcement agents could potentially track innocent people, such as those who happened to be in the vicinity of a crime or disturbance.[23] For example, the ACLU reported in 2010 that Federal Bureau of Investigation agents investigating a series of bank robberies sought the records of every mobile phone that was near each bank when it was robbed. Furthermore, law enforcement agencies access location data frequently, access that could add to concerns about the potential for misuse. For example, in May 2012, Sprint-Nextel reported that it had received over 196,000 court orders for location information over the last 5 years.

Users generally do not know when law enforcement agencies access their location data. In addition to information related to a crime, the location data collected by law enforcement may reveal potentially sensitive destinations, such as medical clinics, religious institutions, courts, political rallies, or union meetings.

Private Sector Entities Have Not Consistently Implemented Recommended Practices to Protect Consumers' Location Privacy

Mobile industry associations and privacy advocacy organizations have recommended practices for industry to better protect consumers' privacy while making use of customers' personal information. Companies we examined have developed privacy policies to disclose information to consumers about the collection of location data and other personal information, but have not consistently or clearly disclosed to consumers what the companies are doing with these data or which third parties they may share them with.

Recommended Practices to Protect Privacy Generally Align with FIPs

Industry associations and privacy advocacy organizations have recommended practices for the mobile industry to better protect consumers' privacy while making use of their personal information. These recommended practices include actions to notify users about the collection and use of their location data, ways users can control data collection, safeguards for user data, and actions to demonstrate accountability. The recommended practices we identified generally align with the FIPs discussed earlier. For example, providing users with controls allowing them to opt in or opt out of having their location data collected aligns with the FIP principles of collection limitation, use limitation, and individual participation, since such controls allow users to limit the collection and use of their personal information while providing them greater ability to be informed about and control how their data are used. Specific examples of recommended practices are shown in table 2.

Table 2. Recommended Location Data Privacy Practices and Alignment with FIPs

Practice category	Examples of specific practices	Alignment with FIPs
Disclosures to users about data collection, use, and sharing	• State reasons companies collect and share data. • State specifically that collection of personal information is limited to specified needs. • Data are not used for a purpose other than what has been disclosed to users without further notice and user consent.	Purpose specification, openness, collection limitation, use limitation
User controls over location data	• Obtain users' consent before collecting their personal information. • Provide users the ability to opt out of data collection they have previously consented to.	Collection limitation, use limitation,individual participation
Data retention and safeguards	• State a specific time frame for retaining user data. • Data should be protected with reasonable security safeguards against risks such as loss or unauthorized access.	Purpose specification, security safeguards, use limitation
Accountability	• Be responsible for protecting users' data.	Accountability

Source: GAO analysis of practices recommended by industry and privacy advocacy organizations.

Companies Take Steps to Protect Privacy, but Not Consistently

Although companies we examined have taken steps to protect the privacy and security of location data, they have not done so consistently, and their actions sometimes fall short of the recommended practices we identified. The 14 mobile industry companies we examined reported actions to inform users about the collection, use, and sharing of their location data primarily through disclosures in their privacy policies.[24]

Companies also disclosed information about ways consumers could control location data collection, how long companies retain location data, how companies safeguard the data, and companies' measures to demonstrate accountability, although how companies addressed these issues varied. While companies' disclosures routinely informed consumers that their location data were being collected, companies' disclosures did not consistently or clearly explain the purposes behind such collection or identify which third parties these data might be shared with.

Data Collection, Use, and Sharing

Recommended practices state that companies should clearly disclose to consumers the collection and use of location data and purpose for doing so. We found that while companies used privacy policies to inform users about location data collection, information about use and sharing was sometimes unclear. All 11 of the mobile carriers, operating system developers, and application developers we examined had privacy policies. Ten of the 11 privacy policies we examined disclosed that the company collected consumers' location data.[25] However, some policies were not clear about how the companies used location data. For example, the privacy policies of 4 of the companies we examined stated ways the companies used "personal information," but did not state whether location data were considered "personal information." It was therefore unclear whether these uses applied to location data. Companies' policies on whether location data were considered personal information varied. Apple's privacy policy, for example, stated that it considered location data to be nonpersonal information.[26] In contrast, T-Mobile's policy stated that location is personally identifiable information. Furthermore, representatives from four of the companies told us that whether location data is considered personal information depends on factors such as how precise the data are and whether they are combined with other information about the user.[27] The operating system developers reported they collected location data in an anonymous manner or took steps to de-identify stored data. In contrast, 3 of the application developers we interviewed stated they stored location data with other personal information about their users.[28] Carriers told us that their practices varied, depending on the specific use of the data.

Recommended practices state that companies should inform consumers about third parties the companies share consumers' data with and the purposes for doing so. Most policies we examined stated the types of third-party companies location data may be shared with, such as application developers and advertisers; however, some policies described third parties with vague terms such as "trusted businesses" or "others." Although some policies stated that the company takes steps to protect this information, such as requiring the third party to follow the company's privacy policy, others made no such statement, and one company's policy said it would not be liable if the third party it shares data with fails to protect it. According to literature examining mobile applications, some applications lack privacy policies and consumers often do not know which companies may receive their personal information after it has been collected by the applications.[29]

Companies also used other methods in addition to privacy policies to inform consumers about location data collection and use, including some methods that informed consumers directly through their phones. For example, some smartphone screens display an icon to indicate when location information is actively being used.

Controls

Recommended practices state that companies should obtain users' consent for collecting, using, and sharing personal information, including location data and explain related controls to users. Companies we contacted reported providing methods for users to control collection and use of location data, but the methods and amount of control varied. Most of these companies indicated that users could control smartphones' use of their location data from the phone; however, the ability to control this varied by operating system, with some providing more options. While all of the operating system developers we examined allowed a user to have location access turned on or off for all applications, some gave users the ability to control whether specific applications could have access to location data. According to the literature we reviewed that examined mobile applications, controls within applications, if available, were sometimes difficult to find. Mobile carriers told us that they do not allow users to control collection of location data for providing basic phone service, since having location data is necessary to provide that service.

All the companies we examined that collected data for providing location-based services indicated that users must first provide consent before location-based services use their location; however, privacy policies we examined did not always explain how users' consent is obtained. Companies told us that a smartphone seeks permission from the user to use location when the user installs an application that makes use of location or the first time the user activates such an application. For example, the iPhone iOS operating system displays a pop-up window the first time a user activates a new application that includes location-based services. The pop-up states that the application is seeking to use the user's location and allows the user to accept or decline. Similarly, Android smartphones notify users that an application will use location at the time a user downloads a new application and seeks user consent through this process.

Retention and Safeguards

The recommended practices we reviewed state that companies should not keep personal information such as location data longer than needed, and some organizations encouraged companies to state a specific data retention time frame. However, 7 of the 11 privacy policies we reviewed did not include a statement about how long the company kept location data. Officials from most companies told us they kept location data only as long as needed for a specific purpose; however, in some cases, this could mean keeping location data indefinitely. The carriers we interviewed named specific time periods for location data retention, which they said varied depending on the specific uses of the data, and reported a range of time from a few days to 3 years after the duration of time a user is a customer with the company. Three companies indicated they kept location data indefinitely, and representatives from one company said they had not established a retention time period. Privacy advocates raised data retention as a particular concern, since the longer companies retain location data, the more likely the potential for misuse. Similarly, FTC's March 2012 report on protecting consumers' private information stated that companies should delete location data as soon as possible, consistent with the services they provide to consumers.[30]

Recommended practices consistently stated the need for companies to safeguard collected user data. Companies reported actions to safeguard users' location data, but practices for how data are safeguarded varied. All the companies we examined reported ways they safeguard users' personal information. For example, all of the privacy policies stated that companies had general security measures in place to protect personal information against loss, theft, or misuse. Specific practices reported by some companies included data encryption, erecting firewalls, and restricting employee access.[31] In some cases, however, it was not clear whether these protections covered location data. As stated above, some privacy policies did not state whether location was considered a form of personal information, and thus it was unclear whether stated safeguards for personal information applied to location data.

Accountability

Most of the recommended practices expressed the need for companies to demonstrate accountability for their practices. However, companies' privacy policies reported few, if any, specific measures for accountability. Five of the 11 privacy policies included general statements that employees were accountable for following the company's policies as outlined in the privacy policy.

A few privacy policies also mentioned that the company followed recommended practices; one carrier's policy stated the company followed recommended practices developed by CTIA-The Wireless Association (CTIA), a nonprofit organization representing mobile carriers and other wireless companies, and 3 companies' policies stated their privacy practices had been certified by TRUSTe, a company that helps companies address privacy issues. Three of the carriers also told us they use their contracts with third parties they share users' personal data with to require those third parties to adhere to CTIA recommended practices for location data. Operating system developers reported varying steps to encourage or require developers of applications that run on their systems to inform users and obtain consent before using their location data.

For example, in 2011, Apple stated that it would reject applications from its on-line store that do not obtain consent from the user before collecting, transmitting, or using a user's location data and that such use must be directly relevant to the features and services provided by the application. In contrast, Google stated that it does not control the behavior of third-party applications in handling location data, but encourages the developers to follow common privacy practices, such as giving users a choice regarding data collection and collecting only necessary information.

Inconsistent Practices to Protect Location Privacy Raise Risks

Companies' inconsistent adherence to recommended practices increases the likelihood that users could be exposed to the privacy risks we discussed previously. For example, because companies have not made clear and consistent disclosures about how they use and share location data, consumers may be unaware which third parties are using their location data (or that third parties are using it at all) and that law enforcement may obtain their location data and use it for surveillance.

Furthermore, because consumers are expected to rely on these disclosures when judging whether they should give consent to a company to access their location, consumers may be providing such consent without complete knowledge of how their data will be used. For example, although privacy policies generally discussed that users' data could be shared with third parties, they sometimes included vague statements like "trusted business partners" rather than specifying the types of companies they shared the data with and the reasons for doing so. Consequently, users lack sufficient information to adequately judge whether they should trust those companies with their personal information.

Privacy advocates we spoke to acknowledged that companies have taken some positive steps to protect privacy, but that the current framework of self-regulation is exposing consumers to unnecessary risks. These advocates said that companies are generally disclosing to users that they will collect location data; however, they are not adequately informing consumers about the uses of the data they collect, including with whom they are sharing the data. These advocates also expressed concern about companies retaining location data longer than necessary, which puts the data at increased risk of inappropriate use. Furthermore, they told us the current framework of self-regulation is insufficient to address these concerns because there are no requirements for companies to consistently implement recommended practices to protect privacy. Federal agencies that have examined location-based services have also noted that the benefits from such services come with concerns. For example, FCC, in its 2012 report on location-based services, noted that such services are expected to deliver $700 billion in value to consumers and business users over the next decade.[32]

However, in summarizing views of participants in a 2011 panel discussion, the FCC report noted that panelists found inconsistency in the privacy notices provided by companies and incomplete disclosure of the ways location data are used. Specifically, the report states that while consumers may have clear notice that an application will collect and use data on their location, these data may be subsequently used in ways that are not transparent to consumers or shared with third parties without consumers' consent. FTC, in its report on protecting consumer privacy, noted that the unauthorized disclosure to third parties of sensitive personal information such as precise location data raises privacy concerns resulting from the unanticipated uses of these data.[33]

Federal Agencies Have Taken Actions to Protect Consumer Privacy, but Additional Actions Could Provide Further Protections

Federal agencies that have responsibility for consumer privacy protection or that interact with the mobile industry have taken steps to promote public awareness, such as providing educational outreach and recommending actions aimed at improving consumer privacy.

However, additional actions could be taken to further protect consumers. For example, NTIA has not defined performance goals for its proposed multistakeholder process, which consists of different groups involved with consumer privacy coming together to discuss relevant issues with the goal of developing codes of conduct for consumer privacy. Additionally, FTC has not issued comprehensive guidance to mobile industry companies with regard to actions they should take to protect mobile location data privacy.

Agencies Have Taken Actions to Promote Awareness of Privacy Issues

Several federal agencies that interact with the mobile industry or have responsibilities for consumer privacy protection have provided educational outreach to the public, developed reports with recommendations aimed at protecting consumer privacy, developed regulatory standards that address mobile-location data privacy, and developed guidance for law enforcement on obtaining mobile location data.

Educational Outreach

FCC and FTC have held educational outreach events, and FTC has developed a fact sheet to educate the public on various privacy issues related to location-based services.

In June 2011, the agencies collaborated to hold a public education forum that explored how consumers can be both knowledgeable and secure when utilizing location-based services. Participants in the forum included representatives from mobile carriers, technology companies, consumer advocacy groups, and academia. Specific topics discussed included

- how location-based services work;
- trends, benefits, and risks of location-based services;
- industry recommended practices; and
- what parents should know about location tracking when their children use mobile devices.

Also in June 2011, FTC issued an informational fact sheet that provided basic information on mobile applications and answered questions on privacy, advertising, and security concerns.[34] Specific topics included the types of data that applications can access on users' devices, the reasons a user's phone collects location data, and ways that applications can cause harm to a user's phone.

In May 2012, FTC held a public workshop on advertising and privacy disclosures to discuss the need for new guidance for online advertisers about making disclosures. Participants included consumer advocates, representatives of industry groups, and academics. The workshop covered topics including

- when, where, and how required disclosures should be made;
- the techniques to increase or decrease the likelihood that consumers will actually read a required disclosure;
- the challenges and best approaches to making adequate disclosures given the screen size constraints of mobile devices; and
- the steps companies can take to communicate with consumers in a clear and consistent way about the companies' privacy practices.

In August 2012, FTC issued guidance for application developers to help developers comply with truth-in-advertising standards and basic privacy principles. The guidance discusses the need for developers to be clear to users about companies' practices to collect and share data, to offer users ways to control how their personal information is collected and shared, and the need to keep users' data secure, among other issues.[35]

Reports on Consumer Privacy

Several agencies have issued or prepared reports that offered recommendations aimed at improving consumer privacy, including location-based services. In February 2012, NTIA prepared a report for the White House on protecting privacy and promoting innovation in the global digital economy.[36] The report offered a framework and expectations for companies that use personal data. The framework includes a consumer privacy bill of rights, a multistakeholder process to specify how the principles in the bill of rights apply in particular business contexts, and effective enforcement. The report also urged Congress to pass consumer data privacy legislation that would, among other things, codify the consumer privacy bill of rights described in the report, grant FTC authority to enforce the bill of rights, and create a national standard under which companies must notify consumers of unauthorized disclosures of certain kinds of personal data.

Also in February 2012, FTC issued a report on privacy disclosures for mobile applications aimed at children. [37] This report highlighted the lack of information available to parents prior to downloading mobile applications for their children and called on the mobile industry to provide greater transparency about their data practices. The report recommended, among other things, that all companies that are involved in developing children's applications—the application stores, developers, and third parties providing services within the applications—should play an active role in providing key information to parents who download applications through simple, short disclosures that are easy to find and understand on the small screen of a mobile device.

In March 2012, FTC issued another report that laid out recommendations for businesses and policy makers aimed at protecting consumer privacy.[38] The report described recommended practices for companies that collect and use consumer data to develop and maintain processes and systems to implement privacy and data security practices. These practices include promoting consumer privacy at every stage of the development of products and services, and giving consumers greater control over the collection and use of their personal data through simplified choices and increased transparency. The report also included recommendations to companies that make use of precise mobile location data, including that they should obtain affirmative express consent from consumers before collecting precise location data; limit collection to data needed for a requested service or transaction; establish standards that address data collection, transfer, use, and disposal, particularly for location data; and, to the extent that location data are collected and shared with third parties, work to provide consumers with more prominent notice and choices about such practices. The report also called on Congress to consider enacting baseline privacy legislation, reiterated FTC's call for legislation governing data security and data broker issues, and urged the industry to accelerate the pace of self-regulation.

Lastly, in May 2012, FCC issued a report that gave an overview of the opportunities and considerations of location-based services. [39] The report describes the value that location-based services may provide to the economy and consumers, challenges companies face as they attempt to provide consumers with appropriate notice and choice, steps the industry is taking to respond to these challenges, and new issues that continue to emerge in this area.

Regulatory Actions

There have been three relevant regulatory actions in the area of protecting mobile location data. In 1998, FCC, implementing requirements of section 222 of the Communications Act, as amended, developed rules to protect CPNI; subsequently, the law was amended to clarify that CPNI includes subscribers' call location data that carriers use to provide telecommunications services.[40] As previously discussed, FCC's regulations limit instances where CPNI can be used or disclosed without customer consent. In November 2000, CTIA proposed the adoption of location information privacy principles that covered the issues of notice, consent, security and integrity of information, and technology neutrality and urged FCC to conduct a rulemaking separate from its general CPNI proceeding, based on CTIA's assessment that the location privacy question is uniquely a wireless concern. In July 2002, FCC declined to initiate a rulemaking because it opined that the amendments to the Communications Act imposed protections for consumers, such as requiring express approval before carriers can use consumers' location information.[41] The Commission decided that rules would be unnecessary and potentially counterproductive because of the still-developing market for location-based services and that CTIA's proposed privacy principles could be adopted by mobile industry companies on a voluntary basis.

In September 2011, FTC proposed amending its rule pertaining to the Children's Online Privacy Protection Act that would revise the definition of personal information to explicitly include location data. 42 According to FTC officials, there is no time frame for the issuance of a final rule in this proceeding, as the Commission is still in the process of evaluating comments.[43]

In June 2012, FCC solicited comments regarding the privacy and data security practices of mobile wireless service providers with respect to customer information stored on their users' mobile communications devices, which could include location information, and the application of existing privacy and security requirements to that information.[44] Since the Commission last solicited public input on this question 5 years ago and technologies and business practices in this area have changed, the Commission sought comments on a variety of issues including:

- the applicability and significance of telecommunications carriers' duty under section 222(a) of the Communications Act to protect customer information stored on their users' mobile communications devices;
- whether the definition of CPNI could apply to information collected at a carrier's direction even before it has been transmitted to the carrier;
- what factors are relevant to assessing a wireless provider's obligations under section 222 of the Communications Act, as amended, and the Commission's implementing rules, or other provisions of law within the Commission's jurisdiction, and in what ways;
- what privacy and security obligations should apply to customer information that service providers cause to be collected by and stored on mobile communications devices; and
- what should be the obligations when service providers use a third party to collect, store, host, or analyze such data.

Guidance for Law Enforcement on Obtaining Mobile Location Data

Justice has developed guidance on how law enforcement may obtain mobile location data,[45] which is primarily obtained through various court orders. These methods have been the subject of recent litigation. There are various methods in which mobile location data can be obtained, including, but not limited to:

- *Warrant*: A warrant allows law enforcement to obtain prospective mobile location data generated by GPS or similar technologies (i.e., where the device is currently located).[46] To obtain a warrant for these data, the government must establish probable cause to believe that the data sought will aid in a particular apprehension or conviction. This method requires the highest standard of evidence of all methods outlined below.
- *Section 2703(d) Court Order*: A 2703(d) court order allows law enforcement officials to obtain certain kinds of historical mobile location data (i.e., where the device was located in the past) that providers collect for business purposes.[47] To obtain this order, the

government must offer specific and articulable facts showing that there are reasonable grounds to believe that the data are relevant and material to an ongoing criminal investigation.

- *Hybrid Order*: Justice has routinely acquired, since at least 2005, certain categories of prospective mobile location data generated by cell tower information through the combination of two court orders, the Pen/Trap court order[48] and the 2703(d) order. The combination order is known as a "hybrid order." To obtain this order, law enforcement officials must affirm that the information likely to be obtained is relevant to an ongoing criminal investigation and further demonstrate specific and articulable facts showing that there are reasonable grounds to believe that the information sought is relevant and material to an ongoing criminal investigation. This order is used because the Communications Assistance for Law Enforcement Act of 1994 precludes law enforcement officials from relying solely on the authority of the Pen/Trap statute to obtain cell tower data for a mobile customer.[49]
- *Section 2702 Voluntary Disclosure*: Communications providers are permitted by law to voluntarily disclose information to law enforcement if the provider, in good faith, believes that an emergency involving danger of death or serious physical injury to any person requires disclosure without delay of communications relating to the emergency.[50]

As already described, law enforcement agencies access location data frequently using these various authorities. Law enforcement's use of location information has spurred courts to review government actions to compel third parties to disclose location data, as judges question and examine what legal standards govern law enforcement access to historical and prospective location information.[51] For example, in 2010, a federal district court in Texas denied government applications for historical cell site data, declaring that compelled warrantless disclosure of cell site data violates the Fourth Amendment.[52] In contrast, in 2012, a federal district court in Maryland upheld the government's use of historical cell site data, concluding that the privacy issues surrounding the collection of historical cell site location records are best left for Congress to decide.[53] Concerns have been raised by privacy advocacy groups about the methods law enforcement can use to obtain location data. For example, the ACLU has opined that existing privacy laws fail to provide adequate legal protections for the increasingly detailed information that is collected by location-based services about consumers' physical locations and that consumers, location-based service providers, and the government are thus acting in uncertain legal territory. Further, most of the privacy advocates we spoke to opined that the government should obtain a warrant based on probable cause of a crime before it tracks, prospectively or historically, the location of a mobile phone or other mobile communications device. This approach seeks to treat historical and prospective location information equally and would require law enforcement to meet a higher standard before obtaining access to any location data.[54]

NTIA's Proposed Stakeholder Process Lacks Defined Performance Goals and an Adequate Enforcement Mechanism

Our Standards for Internal Control in the Federal Government,[55] in conjunction with the Government Performance and Results Act of 1993,[56] state that agencies should set performance goals with specific timelines and measures for program performance. These documents assert that in order to better articulate a results orientation, agencies should create a set of performance goals and measures that addresses important dimensions of performance. They also assert that agencies should use intermediate goals and measures to show progress or contribution to intended results, while including explanatory information on the goals and measures.

Following the February 2012 report on consumer privacy, NTIA began implementing a multistakeholder process, which includes, among other groups, individual companies, industry groups, privacy advocates, and consumer groups. The purpose of the process is to develop codes of conduct that implement the general privacy principles presented in the report and that would be enforceable by FTC if the codes are publicly and affirmatively adopted by mobile industry companies. NTIA believes that the proposed process can provide the flexibility, speed, and decentralization necessary to address policy challenges by facilitating participants' working together to find creative solutions. NTIA also stated that another key advantage of the multistakeholder process is that it can produce solutions in a more timely fashion than a regulatory process. NTIA officials stated that because they are in the beginning stages of defining what the overall process would entail, they could not provide specific information about procedures, deliverables, or time frames. The first session was held on July 12, 2012, and addressed how companies providing applications and interactive services for mobile devices can be transparent about how the companies handle personal data. Officials stated that since the sessions will be driven by the stakeholders, they were unsure if the sessions would cover location data; however, in its comments responding to a draft of this report, NTIA stated that it appears likely stakeholders will address transparency of mobile location-based services based on the topic of conversation at the July meeting. NTIA officials said they planned to hold further discussion sessions, where stakeholders would meet to address distinct issues, but all of the topics have not yet been identified and would be based on recommendations from the stakeholders. Officials stated there is no defined timeline for the remaining discussion sessions or the development of the guiding principles, although in August 2012, NTIA indicated that seven meetings had been scheduled before the end of 2012. Lacking defined performance goals, milestones, and deliverables, it is unclear whether NTIA's multistakeholder process will establish an effective means for addressing mobile location data privacy issues. NTIA officials stated that individual companies' compliance with the codes of conduct produced through the process would be voluntary and that it is uncertain that the process will yield company self-regulations or a third-party monitored code. If companies do not volunteer to follow any resulting principles, enforcement would depend on whether a company's failure to adhere to the agreed-upon practices could be considered an unfair practice. As such, the proposed process does not include any mechanism for enforcing compliance with the guiding principles that may be developed, and NTIA cannot offer any assurance that the results of the process will lead to significant adoption of these principles.

FTC has the authority to take legal action against a company that engages in unfair acts affecting commerce, such as companies engaging in unfair business practices that are likely to cause substantial injury to consumers, which are not reasonably avoidable by consumers themselves. FTC has begun to address mobile location issues by holding public workshops and by releasing a report that laid out recommendations aimed at protecting consumer privacy. It has also developed some guidance for companies that collect, use, and share mobile location data, such as including recommendations on location data collection in its March 2012 consumer data privacy report, including recommendations on improving disclosures to parents about the collection and use of personal information by applications geared toward children in its February 2012 report on that subject, and issuing guidance for application developers regarding collection and use of location data in August 2012. While these various guidelines touch on a number of issues related to mobile location data privacy, FTC has not published comprehensive industry guidance on its views of appropriate actions by mobile companies with regard to privacy. Specifically, by publishing an industry guide for these companies, FTC could help clarify for mobile companies its views on the appropriate actions for protecting privacy of consumers' location data.[57] Doing so could help set expectations for industry on appropriate steps to protect consumers' privacy if the issue has not been adequately addressed through the development and adoption of industry codes or the enactment of legislation. Such guidance could also clarify for companies circumstances under which FTC might take enforcement action against unfair acts.

CONCLUSION

The use and sharing of mobile location data offer benefits to mobile industry companies and consumers, such as providing and improving services and increasing advertising revenue. Nonetheless, these activities can also pose several risks to privacy, including disclosing data to unknown third parties for unspecified uses, consumer tracking, identity theft, threats to personal safety, and surveillance. While mobile industry associations and privacy advocacy organizations have recommended practices for industry to better protect consumers' privacy while making use of customers' personal information, these practices are not mandatory for the companies to implement. Mobile industry companies we examined have inconsistently implemented these practices. In particular, the lack of clear disclosures to consumers about how their location data are used and shared means that consumers lack adequate information to provide informed consent about the use of these data.

Consumers are therefore unable to adequately judge whether the companies with which their data are shared are putting their privacy at risk.

A key federal effort to address these privacy risks is NTIA's planned multistakeholder process, which seeks to develop industry codes of conduct. However, NTIA has not defined the effort's performance goals, milestones, or deliverables. It is therefore unclear if this process will address the risks to privacy associated with the use and sharing of mobile location data. While NTIA recommended that FTC should be granted the authority to enforce any industry codes of conduct that are developed from the multistakeholder process, the current process relies on the industry's voluntary compliance with resulting codes of conduct before FTC could enforce the provisions. Regardless of what results from the multistakeholder process, FTC has authority to take action against companies that engage in unfair and deceptive practices. However, FTC has not issued comprehensive industry guidance establishing its views on the appropriate actions that mobile companies should take to protect consumers' mobile location data privacy. Without clearer expectations for how industry should address location privacy, consumers lack assurance that the aforementioned privacy risks will be sufficiently mitigated.

Recommendations for Executive Action

To address privacy risks associated with the use and sharing of mobile location data, we recommend that the Secretary of Commerce direct NTIA, in consultation with stakeholders in the multistakeholder process, to develop specific goals, time frames, and performance measures for the multistakeholder process to create industry codes of conduct.

To further protect consumer privacy, we recommend that the Chairman of FTC consider issuing industry guidance that establishes FTC's views of the appropriate actions by mobile companies with regard to protecting mobile location data privacy. In developing the guidance, FTC could consider inputs such as industry codes developed through the NTIA multistakeholder process, recommended practices from industry and privacy advocates, and practices implemented by mobile industry companies.

Agency Comments and Our Evaluation

We provided drafts of this report to Commerce, FCC, FTC, and Justice for comment. We also provided relevant portions of the draft to mobile industry companies for comment. We received technical clarifications from all of the agencies and some of the companies, which we incorporated into the report as appropriate. FCC and Justice did not provide comments on the draft.

Commerce provided written comments on a draft of this report. The department disagreed with our recommendation to develop specific goals, time frames, and performance measures for the multistakeholder process to create industry codes of conduct to address privacy risks associated with the use and sharing of mobile phone location data. Specifically, Commerce's letter stated that while NTIA worked with stakeholders to establish a framework that encourages meaningful progress, it is not the agency's role to dictate timelines and deliverables, and that to do so could be counterproductive. We continue to believe that setting goals and time frames for the process could provide stakeholders and consumers with better assurance that the process will indeed result in the timely creation of industry codes to address privacy issues, as called for in the report on consumer privacy that NTIA prepared and that was released by the White House in February 2012. Furthermore, in its letter, Commerce acknowledged NTIA's role in setting a date and selecting a topic for the first multistakeholder process convened in July 2012 and a second process planned to begin in the fall. Thus, we believe it is reasonable to suggest that within its role to initiate and facilitate these meetings, NTIA could work with stakeholders to prioritize consideration of mobile phone location data privacy so that this issue, which, as we previously discussed, has been identified as a particular area of concern by privacy advocates and government agencies, is addressed in a timely manner. We have also revised the wording of the recommendation to state that NTIA's efforts should be done in consultation with the appropriate stakeholders involved in the multistakeholder process to develop industry codes of conduct. FTC provided written comments on a draft of this report, which appear in appendix III. In its letter, FTC stated that it agreed that additional guidance for industry on mobile location data practices would be useful and stated that the agency will continue efforts to inform and guide the industry on best practices for mobile location data. However, FTC also raised concerns with our draft recommendation calling for such guidance to help inform mobile companies how FTC would enforce the prohibition against unfair acts pursuant to the Commission's authority under the Federal Trade Commission Act to take enforcement action against a company that engages in unfair acts affecting commerce. FTC stated that what constitutes unfair facts or practices is determined by statute and the test for determining what is an unfair practice is inherently fact specific in an area in which technology is changing rapidly. It concluded, therefore, that its business guidance efforts may not necessarily be tied to determinations of what is unfair. Consequently, we modified the wording of our recommendation to FTC to focus on the need for FTC to clarify for mobile industry companies its views on appropriate actions companies should take to protect mobile location data privacy.

Sincerely yours,
Mark L. Goldstein
Director, Physical Infrastructure
Gregory C. Wilshusen
Director, Information Security Issues

Appendix I: Objectives, Scope, and Methodology

Our objectives were to examine

1. how mobile industry companies collect location data, why they use and share these data, and how this affects consumers;
2. the types of actions private sector entities have taken to protect consumers' privacy and ensure security of location data; and
3. the actions federal agencies have taken to protect consumer privacy and what additional federal efforts, if any, are needed.

To address all of the objectives, we examined the practices of mobile industry companies involved in the collection and use of location data; specifically mobile carriers, operating system developers, smartphone manufacturers, and application developers. We selected the carriers, operating system developers, and manufacturers with the largest market shares in the United States and the application developers using data on the most popular applications for the two operating systems with the largest market share, Apple iOS and Google Android. See table 3 below for a list of the companies we examined.

Table 3. Mobile Industry Companies We Examined

Type of company	Company names
Mobile carrier	AT&T, Sprint-Nextel, T-Mobile, Verizon
Operating system developer (name of system)	Apple (iPhone iOS), Google (Android), Research in Motion (BlackBerry)
Smartphone manufacturer	Apple, HTC, Motorola, Research in Motion, Samsung
Application developer (sample applications)	Facebook, Google (Gmail, Maps, News andWeather, Search, Talk, YouTube), Pandora (Radio), Rovio Entertainment Ltd. (Angry Birds), Yahoo! (Weather, Messenger)

Source: GAO.

We reviewed and analyzed selected companies' privacy policies and other publicly available documents. We also interviewed representatives of these companies, except Motorola and Samsung, which provided written answers to our questions, and Apple, which declined to answer our questions.

To address our first objective, we reviewed and analyzed relevant literature to determine the various methods companies use to collect location data, why they use and share these data, the benefits that are provided to the consumer, and the associated privacy risks. In addition, we interviewed representatives from mobile industry associations (CTIA – The Wireless Association and Mobile Marketing Association), privacy advocacy groups (American Civil Liberties Union, Center for Democracy and Technology, Electronic Frontier Foundation, Electronic Privacy Information Center, and Future of Privacy Forum), and two privacy researchers (Christopher Soghoian and Ashkan Soltani) who had either testified on the subject before Congress or authored relevant literature on the subject, to discuss the benefits and privacy risks associated with the use of location data. We also interviewed officials from federal agencies that interact with the mobile industry or have responsibilities for consumer privacy protection, including the Federal Communications Commission (FCC), Federal Trade Commission (FTC), Department of Commerce's National Telecommunications and Information Administration (NTIA), and Department of Justice (Justice), to obtain their views.

To address our second objective, in addition to examining the companies as previously discussed, we identified practices recommended by mobile industry associations and privacy advocacy groups to protect the privacy of and secure users' personal information and assessed the extent to which they are consistent with the Fair Information Practices. In addition, we reviewed and analyzed the privacy policies of the selected mobile industry companies to determine their specific practices to protect consumer privacy and how their stated practices aligned with recommended practices. We also reviewed relevant studies of mobile application privacy to obtain further information on how mobile application developers protect consumer privacy. We also interviewed representatives of privacy advocacy groups to obtain their views about how the private sector is protecting users' location privacy.

To address our third objective, we identified and reviewed relevant laws applicable to the mobile industry's use of personal information. To evaluate how federal agencies have ensured compliance with relevant laws and what additional efforts they could take to further protect consumers, we analyzed information and interviewed officials from FCC, FTC, NTIA, and Justice about their enforcement, regulatory, and policymaking efforts to protect consumer privacy. We also interviewed representatives from mobile industry associations and privacy advocacy groups as well as privacy researchers to obtain their views about whether more could be done to protect consumer privacy. In considering ways to address location data privacy issues, we are reporting actions federal agencies could take, rather than potential legislative options.

We conducted this performance audit from December 2011 to September 2012, in accordance with generally accepted government auditing standards. Those standards require that we plan and perform the audit to obtain sufficient, appropriate evidence to provide a reasonable basis for our findings and conclusions based on our audit objectives. We believe that the evidence obtained provides a reasonable basis for our findings and conclusions based on our audit objectives.

End Notes

[1] This estimate includes people with multiple subscriptions. Federal Communications Commission, Annual Report and Analysis of Competitive Market Conditions With Respect to Mobile Wireless, Including Commercial Mobile Services, Fifteenth Report (June 27, 2011).

[2] This estimate is based on a survey conducted by the Centers for Disease Control and Prevention's National Center for Health Statistics. The survey, which seeks to collect information on health issues, also includes questions about household telephones and whether anyone in the household has a wireless phone. S.J. Blumberg and J.V. Luke, Wireless Substitution: Early Release of Estimates from the National Health Interview Survey, July-December 2011, a report for the Centers for Disease Control and Prevention, National Center for Health Statistics (June 2012). Available from: http://www.cdc.gov /nchs/nhis.htm (accessed July 11, 2012).

[3] The Nielsen Company, "Two Thirds of New Mobile Buyers Now Opting For Smartphones" (July 12, 2012).

[4] Kathryn Zickuhr, Three-Quarters of Smartphone Owners Use Location-Based Services, a report for Pew Research Center (Washington, D.C.,: May 11, 2012).

[5] Operating system developers are companies such as Apple, Google, and Research in Motion that develop the software that manages the hardware and applications of smartphones.

[6] See, e.g., Commercial Privacy Bill of Rights Act of 2011, S. 799, 112th Cong. (2011); Electronic Communications Privacy Act Amendments of 2011, S. 1011, 112th Cong. (2011); Geolocational Privacy and Surveillance Act, S. 1212, 112th Cong. (2011); Location Privacy Protection Act of 2011, S. 1223, 112th Cong. (2011); and Geolocational Privacy and Surveillance Act, H.R. 2168, 112th Cong. (2011).

[7] Concurrent with this review, we have examined mobile device security issues, including efforts by manufacturers, which is the subject of a separate report.

[8] A mobile carrier must know the location of a mobile phone to deliver incoming calls.

[9] See U.S. Department of Health, Education, and Welfare, Records, Computers, and the Rights of Citizens: Report of the Secretary's Advisory Committee on Automated Personal Data Systems (Washington, D.C.: July 1973).

[10] Organisation for Economic Co-operation and Development, Guidelines on the Protection of Privacy and Transborder Flow of Personal Data (Sept. 23, 1980). OECD plays a prominent

role in fostering good governance in the public service and in corporate activity among its 30 member countries. It produces internationally agreed-upon instruments, decisions, and recommendations to promote rules in areas where multilateral agreement is necessary for individual countries to make progress in the global economy.

[11] An act or practice is unfair if the injury it causes or is likely to cause to consumers is: (1) substantial; (2) not outweighed by countervailing benefits to consumers or to competition; and (3) not reasonably avoidable by consumers themselves. 15 U.S.C. § 45. A representation, omission, or practice is deceptive if: (1) it is likely to mislead consumers acting reasonably under the circumstances; and (2) it is material, that is, likely to affect consumers' conduct or decisions with respect to the product at issue. See e.g., Federal Trade Commission v. Patriot Alcohol Testers, Inc., 798 F. Supp. 851 (D. Mass. 1992).

[12] Pub. L. No. 105-277, tit. XIII (Oct. 21, 1998, codified at 15 U.S.C. § §6501-6505). In response, FTC promulgated regulations implementing the Children's Online Privacy Protection Act of 1998. See 16 C.F.R. Part 312. Other laws FTC enforces that apply to specific industries could also be applicable to the extent that those companies make use of mobile location data. For example, the Gramm-Leach Bliley Financial Modernization Act requires the establishment of standards for regulated financial institutions to protect customers' privacy and safeguard their personal information, as well as disclose information-sharing practices to consumers and allow customers to limit the sharing of such information with third parties.

[13] CPNI includes information that relates to the quantity, technical configuration, type, destination, location, and amount of use of a telecommunications service as well as information contained in the bills pertaining to telephone service. As the Communications Act requirements for CPNI apply only to carriers, they would not apply to other types of companies that collect and use mobile phone location data, such as application developers. 47 U.S.C. § 222(f), (h).

[14] 47 U.S.C. §222(f)(1).

[15] Pub. L. No. 106-81 (Oct. 26, 1999).

[16] 47 C.F.R. § 20.18.

[17] Pub. L. No. 99-508 (Dec. 21, 1986).

[18] Pub. L. No. 106-81 (Oct. 26, 1999).

[19] The AMBER alert system broadcasts details about local child abductions over area television and radio stations, on highway signage, and, potentially, through other channels. The goal of the system is to enlist the public's help in child recovery efforts.

[20] Personally identifiable information is information that is linked to a specific individual and can be used to locate or identify that person; this information includes an individual's name, aliases, Social Security number, and biometric records.

[21] In the case of mobile carriers, the secondary usage of location data obtained for mobile telephone service is regulated by FCC's CPNI rules, which state that, unless a disclosure is required by law or approved by the customer, telecommunications companies may use, disclose, or permit access to CPNI in order to support their telecommunications services. 47 C.F.R. § 64.2005.

[22] United States Senate, Judiciary Subcommittee on Privacy, Technology and the Law Hearing on Protecting Mobile Privacy: Your Smartphones, Tablets, Cell Phones and Your Privacy, 112th Congress, page 8 (May 10, 2011) (testimony of Ashkan Soltani, Independent Researcher and Consultant).

[23] American Civil Liberties Union of Northern California. Location-Based Services: Time for a Privacy Check-in (San Francisco, Calif.: November 2010).

[24] We reviewed the privacy policies of the carriers, operating system developers, and application developers, which together represented 11 of the 14 companies. We did not review privacy policies for the three manufacturer companies. Some companies represented more than one type of company; for example, Apple is both an operating system developer (iPhone iOS) and manufacturer (iPhone). We sought interviews with all of the companies; 2 of them—Motorola and Samsung—provided written answers to our questions and a third, Apple, did not answer our questions but provided related documents in response to our request.

[25] The privacy policy we examined from Research in Motion, the manufacturer and developer of the operating system for the BlackBerry mobile device, did not mention whether the company collects and uses location data. In a May 9, 2011, letter to Representative Fred Upton, Research in Motion described how BlackBerry devices may collect and use location data, which is explained in a licensing agreement that users must accept. Subsequently, in August 2012, Research in Motion released an updated privacy policy that states the company may collect and store location data associated with the use of location-based services. The policy explains that such data does not personally identify the user.

[26] According to Apple's privacy policy, precise location data are collected anonymously in a form that does not personally identify the user. The policy also states that if Apple combines nonpersonal information with personal information, the combined information is treated as personal information. The policy does not state whether it is Apple's practice to combine location data specifically with personal information.

[27] Four privacy policies stated or implied that personal information collected from users included location data; two policies made it clear that they did not consider location to be personal information; one policy indicated that it varied; and the remaining four policies did not state whether location data was considered personal information.

[28] Two of the other application developers we interviewed told us their applications did not collect location data other than more general data collected during registration such as a user's ZIP code or country. Thus, discussion of whether location data are personally identifiable was not applicable to those companies.

[29] The Future of Privacy Forum examined the ten most popular paid and free applications for three operating systems in September 2011 and found that about half of the applications had privacy policies, and a January 2011 TRUSTe analysis of the top 300 most popular free applications on those three operating systems found that 23 percent had privacy policies, while the others did not.

[30] Federal Trade Commission, Protecting Consumer Privacy in an Era of Rapid Change: Recommendations for Businesses and Policymakers (Washington, D.C.: March 2012).

[31] Manufacturers told us their phones include security features and controls to protect customers' personal information.

[32] Federal Communications Commission, Location-Based Services: An Overview of Opportunities and Other Considerations (Washington, D.C.: May 25, 2012).

[33] Federal Trade Commission, Protecting Consumer Privacy in an Era of Rapid Change: Recommendations for Businesses and Policymakers (Washington, D.C.: March 2012).

[34] The fact sheet is available at http://onguardonline.gov/articles/0018-understandingmobile-apps (accessed Jan. 27, 2012).

[35] The guidance is available at http://business.ftc.gov/documents/bus81-marketing-mobile-app (accessed Aug. 20, 2012).

[36] The White House, Consumer Data Privacy In a Networked World: A Framework for Protecting Privacy and Promoting Innovation in the Global Digital Economy (Washington, D.C.: Feb. 23, 2012).

[37] Federal Trade Commission, Mobile Apps for Kids: Current Privacy Disclosures are Disappointing (Washington, D.C.: February 2012).

[38] Federal Trade Commission, Protecting Consumer Privacy in an Era of Rapid Change: Recommendations for Businesses and Policymakers (Washington, D.C.: March 2012).

[39] Federal Communications Commission, Location-Based Services: An Overview of Opportunities and Other Considerations (Washington, D.C.: May 25, 2012).

[40] The Communications Act of 1934 requires telecommunications providers to secure customers' personal information identified as CPNI. The Wireless Communications and Public Safety Act of 1999 amended section 222 of the Communications Act to make clear that information on the location of customers when they make wireless calls is CPNI and impose specific obligations on carriers not to divulge location information without a customer's explicit consent. 47 U.S.C. § 222.

[41] Specifically, FCC found the amendments to the Communications Act requiring express prior authorization before using or disclosing location information were unambiguous and clearly imposed legal obligations. 47 U.S.C. § 222; FCC, In the Matter of Request by Cellular Telecommunications and Internet Association to Commence Rulemaking to Establish Fair Location Information Practices, FCC 02-208, WT Docket No. 01-72 (July 24, 2002).

[42] The Children's Online Privacy Protection Act of 1998 required FTC to promulgate a rule governing the online collection of information from children under age 13.

[43] 76 Fed. Reg. 59804 (Sept. 27, 2011). In August 2012, FTC issued a supplemental notice of proposed rulemaking pertaining to the Children's Online Privacy Protection Act, 77 Fed. Reg. 46643 (Aug. 6, 2012).

[44] 77 Fed. Reg. 35336 (June 13, 2012).

[45] See, for example, Department of Justice, Executive Office for United States Attorneys. Obtaining and Admitting Electronic Evidence (Washington, D.C.: 2011) and Department of Justice, Computer Crime and Intellectual Property Section Criminal Division. Searching and Seizing Computers and Obtaining Electronic Evidence in Criminal Investigations (Washington, D.C.: 2009).

[46] FED. R. CRIM. P. 41.

[47] 18 U.S.C. §§ 2701-2712.

[48] 18 U.S.C. § 3121. The Pen Register and Trap and Trace Statute allows law enforcement to obtain prospective non-content information associated with communications.

[49] Pub. L. No. 103-414 (Oct. 25, 1994).

[50] 18 U.S.C. § 2702.

[51] Stephanie K. Pell and Christopher Soghoian, "Can You See Me Now?: Toward Reasonable Standards for Law Enforcement Access to Location Data That Congress Could Enact," Berkeley Technology Law Journal, vol. 27, no.1 (2012).

[52] In re U.S. Historical Cell Site Data, 747 F. Supp. 2d 827 (S.D. Tex. 2010). As of our reporting date, an appeal is pending with the U.S. Court of Appeals for the Fifth Circuit.

[53] United States v. Graham, __F. Supp.2d __,2012 WL 691531 (D. Md. 2012).

[54] Pell and Soghoian, "Can You See Me Now?"

[55] GAO, Internal Control: Standards for Internal Control in the Federal Government, GAO/AIMD-00-21.3.1 (Washington, D.C.: November 1999).

[56] Pub. L. No. 103-62 (Aug. 3, 1993).

[57] According to FTC, its industry guides are administrative interpretations by the Commission of the laws it administers and may have application to any matter of fact or law and it may relate to the practices of a particular industry or to practices common to many unrelated industries.

In: Mobile Devices ISBN: 978-1-62618-304-9
Editors: A. R. Walker, D. Brown Miller © 2013 Nova Science Publishers, Inc.

Chapter 2

STATEMENT OF JESSICA RICH, DEPUTY DIRECTOR, BUREAU OF CONSUMER PROTECTION, FEDERAL TRADE COMMISSION. HEARING ON "PROTECTING MOBILE PRIVACY: YOUR SMARTPHONES, TABLETS, CELL PHONES AND YOUR PRIVACY"*

Chairman Franken, Ranking Member Coburn, and members of the Subcommittee, my name is Jessica Rich and I am the Deputy Director of the Bureau of Consumer Protection at the Federal Trade Commission ("FTC" or "Commission").[1] I appreciate this opportunity to appear before you today to discuss the Commission's efforts to protect consumers' privacy in the mobile arena. This testimony first broadly surveys the growth of the mobile marketplace and the Commission's response to this developing industry. Second, it highlights four of the Commission's recent law enforcement actions in the mobile arena, one involving statements that a public relations agency made in the iTunes mobile application store, another involving unsolicited commercial texts, and two recent privacy enforcement actions involving Google and Twitter, major companies in the mobile arena. Finally, it describes

* This is an edited, reformatted and augmented version of a statement presented May 10, 2011 before the Senate Judiciary Committee, Subcommittee on Privacy, Technology and the Law.

the Commission's efforts to address the privacy challenges of these new, and often very personal technologies, including a discussion of how mobile technology is addressed in the privacy framework recently proposed by FTC staff.

I. The Mobile Marketplace

Mobile technology is exploding with a range of new products and services for consumers. According to the wireless telecommunications trade association, CTIA, the wireless penetration rate reached 96 percent in the United States by the end of last year.[2] Also by that same time, 27 percent of U.S. mobile subscribers owned a smartphone,[3] which is a wireless phone with more powerful computing abilities and connectivity than a simple cell phone. Such mobile devices are essentially handheld computers that can not only make telephone calls, but also offer web browsing, e-mail, and a broad range of data services. These new popular mobile devices allow consumers to handle a multitude of tasks in the palm of their hands and offer Internet access virtually anywhere. Companies are increasingly using this new mobile medium to provide enhanced benefits to consumers, whether to provide online services or content or to market other goods or services.[4] Consumers can search mobile web sites to get detailed information about products, or compare prices on products they are about to purchase while standing in the check-out line. Consumers can join texting programs that provide instantaneous product information and mobile coupons at the point of purchase. Consumers can download mobile software applications ("apps") that can perform a range of consumer services such as locating the nearest retail stores, managing shopping lists, tracking family budgets, or calculating tips or debts. Apps also allow consumers to read news articles, play interactive games and connect with family and friends via social media applications. Any of these services can contain advertising, including targeted advertising.

II. FTC's Response to Consumer Protection Issues Involving Mobile Technology

New technology can bring tremendous benefits to consumers, but it also can present new concerns and provide a platform for old frauds to resurface.

Mobile technology is no different. Although there are no special laws applicable to mobile marketing that the FTC enforces, the FTCs core consumer protection law – Section 5 of the FTC Act – prohibits unfair or deceptive practices in the mobile arena.[5] This law applies to marketing in all media, whether traditional print, telephone, television, desktop computer, or mobile device.

For more than a decade, the Commission has explored mobile and wireless issues, starting in 2000 when the agency hosted a two-day workshop studying emerging wireless Internet and data technologies and the privacy, security, and consumer protection issues they raise.[6] In addition, in November 2006, the Commission held a three-day technology forum that prominently featured mobile issues.[7] Shortly thereafter, the Commission hosted two Town Hall meetings to explore the use of radio frequency identification (RFID) technology, and its integration into mobile devices as a contactless payment system.[8] And in 2008, the Commission held a two-day forum examining consumer protection issues in the mobile sphere, including issues relating to ringtones, games, chat services, mobile coupons, and location-based services.[9]

More recently, the agency has invested in new technologies to provide its investigators and attorneys with the necessary tools to monitor and respond to the growth of the mobile marketplace. For example, the Commission has established a mobile technology laboratory, akin to the Commission's longstanding Internet investigative laboratory, containing a variety of smartphones utilizing different platforms and carriers, as well as software and equipment that permit FTC investigators to collect and preserve evidence and conduct research into a wide range of mobile issues, including those related to consumer privacy.

III. Applying the FTC Act to the Mobile Arena

Law enforcement is the Commission's most visible and effective tool for fighting online threats, including those in the mobile marketplace. As described below, the FTC has brought four recent cases that illustrate how Section 5 applies to the mobile arena, including unsolicited text messages and the privacy and security of data collected on mobile devices.

In August 2010, the Commission charged Reverb Communications, Inc., a public relations agency hired to promote video games, with deceptively endorsing mobile gaming applications in the iTunes store.[10] The company allegedly posted positive reviews of gaming apps using account names that gave the impression the reviews had been submitted by disinterested consumers when they were, in actuality, posted by Reverb employees. In addition, the Commission charged that Reverb failed to disclose that it often received a percentage of the sales of each game. The Commission charged that the disguised reviews were deceptive under Section 5, because knowing the connection between the reviewers and the game developers would have been material to consumers reviewing the iTunes posts in deciding whether or not to purchase the games. In settling the allegations, the company agreed to an order prohibiting it from publishing reviews of any products or services unless it discloses a material connection, when one exists, between the company and the product. The *Reverb* settlement demonstrates that the FTC's well-settled truth-in-advertising principles apply to new forms of mobile marketing.

In February, the Commission filed its first law enforcement action against a sender of unsolicited text messages and obtained a temporary restraining order suspending the defendant's challenged operations. The FTC alleged that Philip Flora used 32 pre-paid cell phones to send over 5 million unsolicited text messages – almost a million a week – to the mobile phones of U.S. consumers.[11] Many consumers who received Flora's text messages – which typically advertised questionable mortgage loan modification or debt relief services – had to pay a permessage fee each time they received a message. Many others found that Flora's text messages caused them to exceed the number of messages included in their mobile service plans, thereby causing some consumers to incur additional charges on their monthly bill.[12] The Commission charged Flora with the unfair practice of sending unsolicited text messages and with deceptively claiming an affiliation with the federal government in connection with the loan modification service advertised in the text messages.[13]

The FTC has also taken action against companies that fail to protect the privacy and security of consumer information. Two recent cases highlight the FTC's efforts to challenge deceptive claims that undermine consumers' privacy choices in the mobile marketplace.

First, the Commission's recent case against Google alleges that the company deceived consumers by using information collected from Gmail users to generate and populate a new social network, Google Buzz.[14] The Commission charged that Gmail users' associations with their frequent email contacts became public without the users' consent. As part of the Commission's proposed settlement order, Google must protect the privacy of all of its customers – including mobile users. For example, if Google changes a product or service in a way that makes consumer information more widely available, it must seek affirmative express consent to such a change. This provision applies to any data collected from or about consumers, including mobile data. In addition, the order requires Google to implement a comprehensive privacy program and conduct independent audits every other year for the next 20 years.

Second, in the Commission's case against social networking service Twitter, the FTC charged that serious lapses in the company's data security allowed hackers to obtain unauthorized administrative control of Twitter.[15] As a result, hackers had access to private "tweets" and non-public user information – including users' mobile phone numbers – and took over user accounts, among them, those of then-President-elect Obama and Rupert Murdoch. The Commission's order, which applies to Twitter's collection and use of consumer data, including through mobile devices or applications, prohibits misrepresentations about the extent to which Twitter protects the privacy of communications, requires Twitter to maintain reasonable security, and mandates independent, comprehensive audits of Twitter's security practices.

These are just two recent examples of cases involving mobile privacy issues, but the Commission's enforcement efforts are ongoing.[16] Staff has a number of active investigations into privacy issues associated with mobile devices, including children's privacy.

IV. Mobile Privacy Policy Initiatives

As noted, the rapid growth of mobile technologies has led to the development of many new business models involving mobile services. On the one hand, these innovations provide valuable benefits to both businesses and consumers. On the other hand, they facilitate unprecedented levels of data collection, which are often invisible to consumers.

The Commission recognizes that mobile technology presents unique and heightened privacy and security concerns. In the complicated mobile ecosystem, a single mobile device can facilitate data collection and sharing among many entities, including wireless providers, mobile operating system providers, handset manufacturers, application developers, analytics companies, and advertisers. And, unlike other types of technology, mobile devices are typically personal to the user, almost always carried by the user and switched-on.[17] From capturing consumers' precise location to their interactions with email, social networks, and apps, companies can use a mobile device to collect data over time and "reveal the habits and patterns that mark the distinction between a day in the life and a way of life."[18] Further, the rush of on-the-go use, coupled with the small screens of most mobile devices, makes it even more unlikely that consumers will read detailed privacy disclosures.

In recent months, news reports have highlighted the virtually ubiquitous data collection by smartphones and their apps. Researchers announced that Apple has been collecting geolocation data through its mobile devices over time, and storing unencrypted data files containing this information on consumers' computers and mobile devices.[19] The *Wall Street Journal* has documented numerous companies gaining access to detailed information – such as age, gender, precise location, and the unique identifiers associated with a particular mobile device – that can then be used to track and predict consumers' every move.[20] Not surprisingly, recent surveys indicate that consumers are concerned. For example, a recent Nielsen study found that a majority of smartphone app users worry about their privacy when it comes to sharing their location through a mobile device.[21]

A. Privacy Roundtables

The Commission has been considering these and related issues in connection with its "Exploring Privacy" Roundtable series. In late 2009 and early 2010, the Commission held three roundtables to examine how changes in the marketplace have affected consumer privacy and whether current privacy laws and frameworks have kept pace with these changes.[22] During the second roundtable, one panel in particular focused on the privacy implications of mobile technology, addressing the complexity of data collection through mobile devices; the extent and nature of the data collection, particularly with respect to geolocation data; and the adequacy of privacy disclosures on mobile devices.[23] B. Preliminary Staff Privacy Report Based on the information received through the roundtable process, staff drafted a preliminary report ("Staff Report") proposing a new privacy framework consisting of three main recommendations, each of which is applicable to mobile technology. First, staff recommends [24] that companies should adopt a "privacy by design" approach by building privacy protections into their everyday business practices, such as not collecting or retaining more data than they need to provide a requested service or transaction. Thus, for example, if an app is providing traffic and weather information to a consumer, it does not need to collect call logs or contact lists from the consumer's device. Further, although the app may need location information, the app developer should carefully consider how long the location information should be retained to provide the requested service.

Second, staff recommends that companies should provide simpler and more streamlined privacy choices to consumers. This means that all companies involved in data collection and sharing through mobile devices – carriers, handset manufacturers, operating system providers, app developers, and advertisers – should work together to provide these choices and to ensure that they are understandable and accessible on the small screen. As stated in the Staff Report, companies should also obtain affirmative express consent before collecting or sharing sensitive information such as precise geolocation data.

Third, the Staff Report proposed a number of measures that companies should take to make their data practices more transparent to consumers, including improving disclosures to consumers about information practices. Again, because of the small size of the device, a key question staff posed in the report is how companies can create effective notices and present them on mobile devices.

After releasing the Staff Report, staff received 452 public comments on its proposed framework, a number of which implicate mobile privacy issues specifically.[25] FTC staff is analyzing the comments and will take them in consideration in preparing a final report for release later this year.[26]

CONCLUSION

The Commission is committed to protecting consumers' privacy in the mobile sphere by bringing enforcement where appropriate and by working with industry and consumer groups to develop workable solutions that protect consumers while allowing innovation in this growing marketplace.

End Notes

[1] While the views expressed in this statement represent the views of the Commission, my oral presentation and responses to questions are my own and do not necessarily reflect the views of the Commission or any individual Commissioner.

[2] See CTIA Wireless Quick Facts, available at www.ctia.org/advocacy/research/ index.cfm/ aid/10323.

[3] ComScore, The 2010 Mobile Year in Review Report (Feb. 14. 2011), available at www.comscore.com/Press Events/Presentations Whitepapers/2011/2010 Mobile Year in Rev iew.

[4] Indeed, a recent industry survey found that 62 percent of marketers used some form of mobile marketing for their brands in 2010 and an additional 26 percent reported their intention to begin doing so in 2011. See Vast Majority of Marketers Will Utilize Mobile Marketing and Increase Spending on Mobile Platforms in 2011, ANA Press Release describing the results of a survey conducted by the Association of National Advertisers in partnership with the Mobile Marketing Association, dated January 31, 2011, available at www.ana.net/ content/show/id/20953.

[5] 15 U.S.C. § 45(a).

[6] FTC Workshop, The Mobile Wireless Web, Data Services and Beyond: Emerging Technologies and Consumer Issues, available at www.ftc.gov/bcp/workshops/wireless/ index.shtml.

[7] FTC Workshop, Protecting Consumers in the Next Tech-ade, available at www.ftc.gov/bcp/workshops/techade. The Staff Report is available at www.ftc.gov/os/ 2008/03/P064101tech.pdf.

[8] FTC Workshop, Pay on the Go: Consumers and Contactless Payment, available at www.ftc.gov/bcp/workshops/payonthego/index.shtml; FTC Workshop, Transatlantic RFID Workshop on Consumer Privacy and Data Security, available at www.ftc.gov/bcp/ workshops/transatlantic/index.shtml.

[9] FTC Workshop, Beyond Voice: Mapping the Mobile Marketplace, available at www.ftc.gov/ bcp/workshops/mobilemarket/index.shtml.

[10] Reverb Commc'ns, Inc., FTC Docket No. C-4310 (Nov. 22, 2010) (consent order).

[11] FTC v. Flora, CV11-00299 (C.D. Cal.) (Compl, filed Feb. 22, 2011).

[12] While the financial injury suffered by any consumer may have been small, the aggregate injury was likely quite large. And, even for those consumers with unlimited messaging plans, Flora's unsolicited messages were harassing and annoying, coming at all hours of the day.

[13] The complaint against Flora also alleges violations of the CAN-SPAM Act for sending unsolicited commercial email messages advertising his texting services that did not include a valid opt-out mechanism and failed to include a physical postal address. In these emails, Flora offered to send 100,000 text messages for only $300. See FTC Press Release, FTC Asks Court to Shut Down Text Messaging Spammer (Feb. 23, 2011), available at www.ftc.gov/opa/2011/02/loan.shtm.

[14] Google, Inc., FTC File No. 102 3136 (Mar. 30, 2011) (consent order accepted for public comment).

[15] Twitter, Inc., FTC Docket No. C-4316 (Mar. 2, 2011) (consent order).

[16] See also FTC v. Accusearch, Inc., 2007 WL 4356786 (D. Wyo. Sept. 28, 2007) (operation of a website that illegally obtained telephone records, including cell phone records, through pretexting was an unfair act) ,aff'd, 570 F.3d 1187 (10th Cir. 2009).

[17] See, e.g., Pew Internet & American Life Project, Adults, Cell Phones and Texting at 10 (Sept. 2, 2010), available at www.pewinternet.org/Reports/2010/Cell-Phones-and-American-Adults/Overview.aspx ("65% of adults with cell phones say they have ever slept with their cell phone on or right next to their bed"); Teens and Mobile Phones at 73 (Apr. 20, 2010), available at www.pewinternet.org/Reports/2010/Teens-and-Mobile-Phones/ Chapter-3/Sleeping-with-the-phone-on-or-near-the-bed.aspx (86% of cell-owning teens ages 14 and older have slept with their phones next to them).

[18] United States v. Maynard, 615 F.3d 544, 562 (D.C. Cir. 2010).

[19] See Jennifer Valentino-Devries, Study: iPhone Keeps Tracking Data, WALL ST. J. (Apr. 21, 2011), available at http://online.wsj.com/article/SB10001424052748704570 7045762753 23811369758.html.

[20] See, e.g., Robert Lee Hotz, The Really Smart Phone, WALL ST. J. (Apr. 23, 2011), available at http://online.wsj.com/article/SB10001424052748704547604576263 261679848814. html?mod= (describing how researchers are using mobile data to predict consumers' actions); Scott Thurm & Yukari Iwatane Kane, Your Apps are Watching You, WALL ST. J. (Dec. 18, 2010), available at http://online.wsj.com/article/SB1000142405 2748704 368004576027751867039730.html?mod= (documenting the data collection that occurs through many popular smartphone apps).

[21] NielsenWire, Privacy Please! U.S. Smartphone App Users Concerned with Privacy When It Comes to Location (Apr. 21, 2011), available at http://blog.nielsen.com/ nielsenwire/ online_mobile/privacy-please-u-s-smartphone-app-users-concerned-with-privacy-w hen-it-comes-to-location/; see also Ponemon Institute, Smartphone Security: Survey of U.S. Consumers at 7 (Mar. 2011), available at http://aa-download.avg.com/ filedir/other/ Smartphone.pdf (64% of consumers worry about being tracked when using their smartphones).

[22] See FTC, Exploring Privacy: A Roundtable Series, available at http://www.ftc.gov/bcp/ workshops/privacyroundtables/index.shtml.

[23] Transcript of Roundtable Record, Exploring Privacy: A Roundtable Series at 238 (Jan. 28, 2010) (Panel 4, "Privacy Implication of Mobile Computing), available at http://www. ftc.gov/bcp/workshops/privacyroundtables/PrivacyRoundtableJan2010Transcript. pdf.

[24] See FTC Staff Report, Protecting Consumer Privacy in an Era of Rapid Change: A Proposed Framework for Businesses and Policymakers (Dec. 1, 2010), available at http://ftc.gov/os/2010/12/101201privacyreport.pdf. Commissioners Kovacic and Rosch issued concurring statements available at http://ftc.gov/os/2010/12/101201 privacyreport. pdf at Appendix D and Appendix E, respectively.

[25] See Comment of CTIA (Feb. 18, 2011), available at http://www.ftc.gov/os/comments/ privacyreportframework/00375-58002.pdf; Comment of Verizon and Verizon Wireless (Feb. 18, 2011), available at http://www.ftc.gov/os/comments/ privacyreportframework/ 00428-58044.pdf; see also, e.g., Comment of Center for Digital Democracy and U.S. PIRG at 10-11, 20-21, 33 (Feb. 18, 2011), available at http://www.ftc.gov/os/comments/ privacyreportframework/00338-57839.pdf; Comment of Stanford Security Laboratory at 11-12 (Feb. 18, 2011), available at http://www.ftc.gov/os/comments/ privacy report framework/00467-57980.pdf.

[26] Another major initiative addressing the mobile marketplace is the Commission's review of the Children's Online Privacy Protection Rule, issued pursuant to the Children's Online Privacy Protection Act ("COPPA"). Initiated in April 2010, this review sought public comment on whether technological changes to the online environment warrant any changes to the Rule or to the statute. In June 2010, the Commission also held a public roundtable to discuss the implications for COPPA enforcement raised by new technologies, including the rapid expansion of mobile communications. The Rule review is ongoing.

In: Mobile Devices ISBN: 978-1-62618-304-9
Editors: A. R. Walker, D. Brown Miller © 2013 Nova Science Publishers, Inc.

Chapter 3

STATEMENT OF JUSTIN BROOKMAN, DIRECTOR, CONSUMER PRIVACY, CENTER FOR DEMOCRACY AND TECHNOLOGY. HEARING ON "PROTECTING MOBILE PRIVACY: YOUR SMARTPHONES, TABLETS, CELL PHONES AND YOUR PRIVACY"*

Chairman Franken, Ranking Member Coburn, and Members of the Subcommittee:

On behalf of the Center for Democracy & Technology (CDT), I thank you for the opportunity to testify today. We applaud the Chairman's leadership in examining the privacy issues presented by location-enabled mobile devices and appreciate the opportunity to address the lack of legal protection facing of what is one of the fastest growing areas of technological innovation.

CDT is a non-profit, public interest organization dedicated to preserving and promoting openness, innovation, and freedom on the decentralized Internet. I will briefly note the particular privacy issues presented by mobile services, and then describe the inadequacy of existing law to protect consumers. CDT strongly believes that legislation based on the full range of

* This is an edited, reformatted and augmented version of testimony given on May 10, 2011 before the Senate Judiciary Committee, Subcommittee on Privacy, Technology and the Law.

Fair Information Practice Principles (FIPPs) should be enacted to address the privacy challenges faced in the mobile space.

1. The Promise and Peril of Location-Enabled Mobile Devices

Mobile phones and tablets have exploded in popularity in recent years, and all evidence indicates that this trend will continue. Smartphone sales are expected to eclipse those of desktop and laptop computers combined in the next two years.[1] However, mobile devices store and transmit a particularly personal set of data. These devices typically allow third parties to access personal information such as contact lists, pictures, browsing history, and identifying information more readily than in traditional internet web browsing. The devices also use and transmit information consumer's precise geolocation information as consumers travel from place to place.

At the same time, consumers have less control over their information on mobile devices than through traditional web browsing. While third parties, like ad networks, usually must use "cookies" to track users on the web, they often get access to unique — and unchangeable — unique device identifiers in the mobile space. While cookies can be deleted by savvy users, device identifiers are permanent, meaning data shared about your device can always be correlated with that device. As is the case with most consumer data, information generated by mobile devices is for the most part not protected by current law and may be collected and shared without users' knowledge or consent.

Consumers interact with their mobile devices by running applications, or "apps" (i.e., programs designed to run on mobile devices). The mobile apps ecosystem is robust and offers an ever-increasing range of functionality from games, music, maps, instant messaging, email, metro schedules, and more. Mobile apps may be preinstalled on the device by the manufacturer or distributor, or users can download and install the programs themselves from their operating system's "apps store" (like iTunes or the Android Market), or a third-party store (like Amazon). App developers range from large, multinational corporations to individuals coding in their parents' basements. Generally speaking, we have seen a vibrant and creative app market develop for mobile devices. Unfortunately, it can be hard to know what information these apps have access to and with whom they are sharing it.

Recent studies of this flourishing apps data ecosystem have unearthed troubling findings. A recent survey indicated that of the top 340 free apps, only 19% contained a privacy policy *at all.*[2] Last December, the Wall Street Journal investigated the behavior of the 101 most popular mobile apps, finding that more than half transmitted the user's unique device ID to third parties without the user's consent. [3] Forty-seven apps transmitted the phone's location.[4] One popular music app, Pandora, sent users' age, gender, location and phone identifier to various ad networks.[5] In sum, a small phone can leak a big amount of data.

Once an app has access to a user's data, there are usually no rules governing its disclosure, and no controls available to consumers to regain control of it. For the most part, once data leaves the phone, it is effectively "in the wild." It may be retained long after the moment of collection, and often long after the original service has been provided. App developers, advertisers, ad networks and platforms, analytics companies, and any number of other downstream players can share, sell, or unpredictably use data far into the future. Even insurance companies are eying data mined from online services for new predictive models.[6] In short, today's mobile environment provides a gateway into an opaque and largely unregulated market for personal data.

Location data is of particular concern. In recent years, the accuracy of location data has improved while the expense of calculating and obtaining it has declined. As a result, location-based services are an integral part of users' experiences and an increasingly important market for U.S. companies. Consumers like the convenience and relevance of location based services. Location data can be used guide you to the closest coffee shop or help you navigate an unfamiliar neighborhood. Your location can be leveraged to connect you with coupons or deals in your immediate vicinity. And new, innovative, and useful services are introduced daily.

People generally carry their mobile devices wherever they go, making it possible for location data be collected everywhere, at any time, and potentially without prompting. Understandably, many find the use of location data without clear transparency and control troubling. Research shows that people value their location privacy and are less comfortable sharing their location with strangers than with acquaintances, and want granular control over their location information.[7] Indeed, location data is especially sensitive information that can be used to decipher revealing facts or put people at physical risk. Location information could disclose visits to sensitive destinations, like medical clinics, courts and political rallies. Access to location can also be used in stalking and domestic violence.[8] Finally, as an increasing number of minors carry location-capable cell phones and devices, location privacy may become a child safety matter as well.

There are also questions and concerns about the collection, usage, and storage of data by mobile platform providers such as Apple and Google. Because in many instances, these companies are the ones actually calculating your location (based on comparing the WiFi access points in range of your device with known databases), they may receive extremely detailed information about consumer activity, considerably more so than traditional computer operating systems.

Although these companies typically assert that data they receive from consumers is anonymized and used merely to build out their databases of access points, these limitations are self-imposed. Furthermore, these platforms may store detailed location and other customer information on the phone itself, which could then be accessed by government officials, potentially without a warrant, malicious hackers, or merely the person who finds your lost phone at Starbucks.[9]

Mobile devices and the services they enable provide consumers with great benefit. But it is imperative that Congress provide a clear policy framework to protect users' privacy and trust.

CDT strongly supports privacy legislation that implements the full range of Fair Information Practice Principles (FIPPs) across all consumer data and provides enhanced protections for sensitive information, such as precise geolocation, including enhanced, affirmative opt-in consent. Unfortunately, today's legal protections fall far short.

2. Existing Legal Protections for Mobile Device Information are Outdated, Inapplicable, or Unclear

A number of laws aim to protect electronic communications, including location information. Unfortunately, technology has far outpaced these statutory protections in both the commercial and government contexts. An update is long overdue.

Following is a summary of relevant laws and an analysis of their application to today's location-enabled mobile devices.

A. The Telecommunications Act of 1996 and Cable Communications Policy Act of 1984 (CPNI Rules)

Through the Telecommunications Act of 1996, with subsequent amendments, Congress has prohibited a telecommunications carrier from disclosing customer proprietary network information (CPNI), including "information that relates to the . . . location . . . [of] any customer of a telecommunications carrier . . . that is made available to the carrier by the customer solely by virtue of the carrier-customer relationship" — except in emergency contexts or "as required by law or with the approval of the customer."[10] In short, Congress issued a minimal standard that prohibited carriers from releasing location and other customer information on a solely discretionary basis.

Fifteen years ago, these privacy rules were a groundbreaking development. At the time, telecommunications carriers served as the primary gatekeepers for location information. Data about a cell phone user's location was calculated within a *carrier's* network using signals sent by the phone to the *carrier's* service antennas. These traditional protections have been left behind as we move from voice (traditionally the purview of telecommunications carriers) to data (which is often not the prevue of telecommunications carriers).

In light of modern location technology, there are at least two major shortcomings of the CPNI statute and resulting Federal Communications Commission (FCC) rules:

1) The CPNI rules simply do not apply to new types of location technologies, applications, and services. More specifically, the CPNI rules do not cover methodologies that are independent of telecommunications carriers covered by the law (e.g., WiFi database lookups, cell tower database lookups, or unassisted GPS locations). Thus, when an iPhone or Android user installs a location-based application, the location data transmitted by the resulting service is very likely completely unregulated under the CPNI rules.
2) Even, when a telecommunications carrier is involved in providing a location based service, it may not be covered by the CPNI rules because the FCC has removed wireless broadband service from Title II of the Communications Act (to which the CPNI rules apply) and deregulated it. When the Commission issued its Wireless Broadband Order,[11] Commissioner Copps explained the fractured effect of the

Order on the protection of location information under the CPNI rules.[12]

Thus, modern mobile devices leverage location services that are largely invisible to the telecommunications provider and thus very likely outside the scope of the law. Although Congress and then the FCC did extend CPNI rules to cover IP-enabled ―interconnected‖ VoIP services,[13] that protection still only extends to voice service regulated under Title II. At best, the application of CPNI rules to carrier-provided location-based data services is a murky question; at worst, the CPNI rules provide no protection whatsoever.

Practically speaking, this creates some striking confusion. A consumer using a mobile phone today can be protected by the CPNI rules one moment and unprotected the next. For example, a user might place a phone call using the traditional Commercial Mobile Radio Service (CMRS). In this case, they could feel secure that the CPNI rules required their carrier to protection their information. After the call, they use an Internet-based app or location service that uses location data rendered apart from the telecommunications carrier. Here, the user is likely unprotected.

B. The Electronic Communications Privacy Act (ECPA)

The Electronic Communications Privacy Act was passed in 1986 primarily to address the issue of government access (about which, see below). However, it also contains important limitations on how companies may voluntarily share with other companies customer communications. Most notably, the law prohibits certain companies from sharing the content of customer communications or records without their consent.[14] In theory, this might prohibit mobile operating systems or applications from sharing consumer data without permission. Unfortunately, ECPA, while a very important and forward-looking statute at the time it was passed, was not written with the mobile apps ecosystem in mind. As applied to the current mobile environment, ECPA as a limitation on inter-business sharing of consumer data is, at best, vague and uneven.

When discussing the kinds of mobile applications and services at issue here today, it is not even clear which parties are currently covered by ECPA. ECPA's coverage of stored communications extends only to two categories of services — electronic communications services (ECSs) and remote computing services (RCSs). An ECS is a service that permits users to send or receive communications information (defined in part as ―signs, signals, writing, images, sounds, data, or intelligence of any nature‖)[15] to a third party or parties, like an email service or a private bulletin board such as a restricted Facebook wall. Some apps and location-based services are ECSs, some are not, and some fall into a grey area. For example, a service that allows users to share their location with a specific group of friends or associates is likely an ECS, with the ―data or intelligence‖ communicated to friends being the combination of the user's identity and her location data. However, an app that allows a user to share his location with a restaurant chain solely to allow it to return the location of the nearest restaurant is likely not an ECS, because it does not provide a way to communicate with third parties The statute ultimately requires highly fact-dependent analysis on the ECS question.

Remote computing services are, if anything, even more murky. An RCS includes any service that provides to the public computer storage or processing. The limited case law developed around this definition has not clarified its boundaries. Courts have held that websites enabling certain commercial transactions are not RCSs, but have suggested that remote processing of user-collected or -generated data is likely to be covered. Almost any app that collects user location or personal data and sends it to a remote server for further processing could, theoretically, fall under the ambit of this provision. However, it is important to note that mobile operating systems — the entities that often generate consumer location information in the first place — likely do not qualify as either ECSs or RCSs, and thus ECPA offers no protections at all as to those companies.

Of course, even if an app were to fall under the ECPA's ambit, there would still be open questions about whether customer data constituted the ―content‖ of a communication subject to protection. If a consumer affirmatively sent a location request to an app maker to ask for a nearby bar or restaurant, ECPA could arguably restrict the transfer of that information to third parties because the consumer's location was the content of a customer-initiated communication. If on the other hand, the app accessed the user's location in the background merely in order to send to a third party to serve relevant advertising, such request probably would not be governed. Such a reading of the statute would however lead to the perverse result that a consumer's information is afforded greater protections when she affirmatively shares sensitive data, as opposed to when her data is shared without her knowledge or consent.

Though the issue is not the focus of the present hearing, it is important to note that legislation to clarify the standards for government access to that information should also remain a Congressional priority. While the Communications Assistance for Law Enforcement Act (CALEA) indicates what the standard for law enforcement access to location information *is not,* no statute indicates what the standard for law enforcement access *is.* CALEA provides that a pen register or trap and trace order[16] cannot be used to obtain location information, but that statute is silent on what the standard should be.[17] There is a federal statute on tracking devices, but it does not specify the standard that law enforcement must meet in order to place such a device.[18] Most importantly, the Electronic Communications Privacy Act (ECPA),[19] which sets up the sliding scale of authority for governmental access to information relating to communications (ranging from mere subpoena to warrant), does not specify what standard applies to location information.

This has resulted in a mish-mash of confused decisions while courts struggle to find and apply a legal standard. It has led to sometimes arbitrary distinctions based on whether location information is sought in real time or from storage, the degree of precision in the location information sought, the period(s) during which location information is sought, and the technology used to generate the location information. Some courts[20] have adopted a ―hybrid theory‖ advanced by the Department of Justice, holding that location information is accessible to government *in real time* if it meets the standard for *stored* transactional information in Section 2703(d) of the Stored Communications Act.[21] Other courts have required a higher level of proof – probable cause – for law enforcement access to this prospective location information.[22] As one federal magistrate judge recently testified in front of the House Judiciary Committee, there is no comprehensible standard for magistrate judges to apply when the government requests access to cell site location data – just an incoherent array of competing court decisions.[23]

As the first few circuit court decisions to address governmental requests for location information of all types have started to come down, it is becoming clear that the courts have constitutional concerns with these requests. In August, the D.C. Circuit held that putting a device in place to engage in extended GPS tracking without a warrant violates the Fourth Amendment.[24] In September, the Third Circuit held that magistrate judges faced with a request from the government for cell site location information have discretion under ECPA to insist upon a showing of probable cause, in part because of the potential sensitivity of the information.[25] Both the confusion in the lower courts and the consternation in the appeals courts demonstrate that Congressional attention to these statutes is sorely needed.

Congress enacted ECPA in 1986 to foster new communications technologies by giving users confidence that their privacy would be respected. ECPA helped further the growth of the Internet and proved monumentally important to the U.S. economy. Now, technology is again leaping ahead, but the law is not keeping up. CDT — through its Digital Due Process coalition — has convened technology and communications companies, privacy advocates and academics to create four principles for reforming ECPA for the next quarter-century. One of those principles is that location information should only be accessed through the use of a warrant[26] and we believe Congress should enact legislation that imposes a warrant requirement. Though the larger ECPA reform effort is and should remain independent of the issues being discussed here today, CDT believes setting easily-understood privacy-protective standards for government access to location data is a critical component of ensuring the privacy of American citizens and the success of American technology service providers.

C. The Computer Fraud and Abuse Act (CFAA)

The Computer Fraud and Abuse Act (CFAA) is a criminal statute that prohibits intentional trespass into and theft from protected computer systems.[27] It criminalizes, in relevant part, one who ¯intentionally accesses a computer without authorization or exceeds authorized access . . . information from any protected computer.‖[28] In short, it's a law to prosecute malicious hackers.

The CFAA is a law design to combat egregious computer crimes and cannot, and should not, be a primary tool in protecting consumers' mobile privacy from data sharing for marketing or related purposes. In the past, there have been failed attempts to stretch the CFAA to cover contractual terms of service.[29] CDT has warned that these attempts come with troubling encroachments on civil liberties and freedom of speech.[30] Criminal sanctions for certain computer crimes might well deter bad actors and provide appropriate tools in extreme circumstances. However, it is a blunt instrument not designed to address mobile privacy challenges arising from commercial activity.

The mobile market is nascent and innovating quickly. Many mobile app developers are individuals or small startup companies. They might be amateur programmers, working with various prefabricated pieces of code and advertising solutions. They may or may not have expertise in privacy or relevant law. Criminal sanctions, including jail time, would be heavy-handed and would likely chill the innovation we see today.

D. Federal Trade Commission Act and State Attorneys General

Absent any affirmative legal requirements provided by sectoral specific privacy laws (such as those governing health or financial data), the default privacy rule for most consumer data is set by the FTC Act's prohibition on unfair and deceptive trade practices.[31] Under this authority, the FTC has established some general precedents about what constitutes a deceptive or unfair privacy practice online, such as recent settlements against companies who offered deceptive and ineffective opt-out solutions, and against Google for sharing personal data with other Google customers in violation of previous representations as part of the Buzz product. While these cases are important, they also demonstrate that the FTC is generally limited under current law to bringing enforcement actions against companies that make affirmative misstatements about their own privacy practices. In the absence of a baseline federal privacy law that gives the FTC the tools it needs and establishes it as the lead law enforcement agency for privacy matters, consumer protections in the location privacy space will continue to fall short. State Attorneys General also have consumer protection mandates that allow them to pursue service providers that engage in unfair or deceptive trade practices. To date, however, perhaps due to the inherent limitations in their authority, relatively little attention has been paid at the state level to consumer privacy concerns.

3. The Need for Congressional Action

Given that the default rule for most consumer data — including sensitive location data — is merely that companies cannot make affirmative misstatements about the use of that data, CDT strongly supports the enactment of a uniform set of baseline rules for personal information collected both online and offline. Modern data flows often involve the collection and use of data derived and combined from both online and offline sources, and the rights of consumers and obligations of companies with respect to consumer data should apply to both as well. The mobile device space implicates many different kinds of data in a complicated ecosystem. Cramming more notices onto small screens is alone insufficient. We need a data privacy law that incentivizes and requires companies to provide clear and conspicuous notice to consumers about the use of their information and provides for meaningful control of that information. Moreover, companies should collect only as much personal information as necessary, be clear about with whom they're sharing information, and expunge information after it is no longer needed.

The Fair Information Practices (FIPPs) should be the foundation of any comprehensive privacy framework. FIPPs have been embodied to varying degrees in the Privacy Act, Fair Credit Reporting Act, and other sectoral federal privacy laws that govern commercial uses of information online and offline. The most recent formulation of the FIPPs by the Department of Homeland Security offers a robust set of modernized principles that should serve as the foundation for any discussion of consumer privacy legislation.[32] Those principles are:

- Transparency
- Purpose Specification
- Use Limitation
- Data Minimization
- Data Accuracy
- Individual Participation
- Security
- Accountability

For particularly sensitive data, such as health information, financial information, information about religion or sexuality, and — most relevant here — precise geolocation data, a legislative framework should provide for enhanced application of the Fair Information Practice Principles, including for affirmative opt-in consent for the collection and/or transfer of such information. Consumers understandably have greater concerns about the use and storage of such information, and the law should err against presuming a consumer's assent to share such information with others. Furthermore, as noted above, the laws governing government access to consumer data should be modernized to require a warrant to access sensitive location information.

CONCLUSION

CDT would like to thank the Subcommittee again for holding this important hearing. We believe that Congress has a critical role to play in ensuring the privacy of consumers in the growing market of mobile devices and services. CDT looks forward to working with the Members of the Subcommittee as they pursue these issues further.

End Notes

[1] Cecilia Kang, Smartphone sales to pass computers in 2012: Morgan Stanley analyst Meeker, THE WASHINGTON POST, November 11, 2010, http://voices.washingtonpost.com/posttech/2010/11/smartphone_sales_to_pass_compu.html.

[2] Mark Hachman, Most Mobile Apps Lack Privacy Policies: Study, PC MAGAZINE, April 27, 2011, http://www.pcmag.com/article2/0,2817,2384363,00.asp.

[3] Scott Thurm and Yukari Iwatani Kane, Your Apps are Watching You, THE WALL STREET JOURNAL, December 17, 2010, http://online.wsj.com/article/SB10001424052748704694004576020083703574602.html

[4] Id.

[5] Id.

[6] Leslie Scism and Mark Maremont, Insurers Test Data Profiles to Identify Risky Clients, THE WALL STREET JOURNAL, November 19, 2010, http://online.wsj.com/article/SB10001424052748704648604575620750998072986.html.

[7] See, e.g., Janice Y. Tsai, Patrick Kelley, Paul Drielsma, Lorrie Cranor, Jason Hong, Norman Sadeh, Who's viewed you?: the impact of feedback in a mobile location-sharing application, Conference on Human Factors in Computing Systems: Proceedings of the 27th international conference on human factors in computing systems (2009), http://www.cs.cmu.edu/~sadeh/Publications/Privacy/CHI2009.pdf; Sunny Consolvo, Ian E. Smith, Tara Matthews, Anthony LaMarca, Jason Tabert, and Pauline Powledge, Location Disclosure to Social Relations: Why, When, & What People Want to Share, CHI '05: Proceedings of the SIGCHI conference on human factors in computing systems (2005), www.placelab.org/publications/pubs/chi05- locDisSocRel-proceedings.pdf.

[8] See, e.g., Rob Stafford, Tracing a Stalker, Dateline NBC, June 16, 2007, http://www.msnbc.msn.com/id/19253352/.

[9] See Alexis Madrigal, What Does Your Phone Know About You? More Than You Think, THE ATLANTIC, April 25, 2011, http://www.theatlantic.com/technology/archive/2011/04/what-does-your-phone-know-about-you-more-than-youthink/237786/.

[10] 47 U.S.C. § 222.

[11] Appropriate Regulatory Treatment for Broadband Access to the Internet Over Wireless Networks, Declaratory Ruling, WT Docket No. 07-53, FCC 07-30, 2 (rel. Mar. 23, 2007).

[12] Id at 2 (carriers offering Title I services ―appear[] to be entirely free, under our present rules, to sell off aspects of the customer[s'] call or location information to the highest bidder.‖).

[13] See 47 C.F.R. § 64.2001, et seq.

[14] 18 U.S.C. §§ 2702(a).

[15] 18 U.S.C. §§ 2510(12).

[16] A pen register/trap and trace order permits law enforcement to obtain transactional, non-content information about wire and electronic communications in real time, including numbers dialed on a cellular telephone and telephone numbers of calls coming into a cell phone. See 18 U.S.C. §§ 3121-3127.

[17] 47 U.S.C. § 1002(a)(2).

[18] 18 U.S.C. § 3117.

[19] 18 U.S.C. §§ 2510 et seq.

[20] See, e.g., In re Application of U.S. for an Order for Disclosure of Telecommunications Records and Authorizing the Use of a Pen Register and Trap and Trace, 405 F. Supp. 2d 435 (S.D.N.Y. 2005).

[21] The SCA, part of the Electronic Communications Privacy Act, is codified at 18 U.S.C. §§ 2701 et seq.

[22] See, e.g., In re Application for Pen Register and Trap/Trace Device with Cell Site Location Authority, 396 F. Supp. 2d 747 (S.D.Tex. 2005).

[23] See Electronic Communications Privacy Act Reform and the Revolution in Location Based Technologies and Services Before the H. Comm. on Judiciary Subcomm. on the Constitution, Civil Rights, and Civil Liberties, 111th Cong. (June 24, 2010) (statement of Stephen Wm. Smith, United States Magistrate Judge).

[24] U.S. v. Maynard, 615 F.3d 544 (D.C. Cir. 2010).

[25] In the Matter of the Application of the United States of America for an Order Directing a Provider of Electronic Communication Service to Disclose Records to the Government, 620 F.3d 304 (3d Cir. 2010).

[26] For more information on the Digital Due Process coalition and its principles, see Digital Due Process at http://www.digitaldueprocess.org.

[27] 18 U.S.C. § 1030.

[28] 18 U.S.C. § 1030(a)(2)(C).

[29] See generally, US v. Drew, Electronic Frontier Foundation, available at https://www.eff.org/cases/united-states-vdrew (last visited May 6, 2011).

[30] Id.

[31] The FTC Act, 15 U.S.C. §§ 41 et seq.

[32] U.S. Department of Homeland Security, Privacy Policy Guidance Memorandum, The Fair Information Practice Principles: Framework for Privacy Policy at the Department of Homeland Security, December 2008, http://www.dhs.gov/xlibrary/assets/privacy/privacy_policyguide_2008-01.pdf.

In: Mobile Devices ISBN: 978-1-62618-304-9
Editors: A. R. Walker, D. Brown Miller © 2013 Nova Science Publishers, Inc.

Chapter 4

TESTIMONY OF ASHKAN SOLTANI, INDEPENDENT RESEARCHER AND CONSULTANT. HEARING ON "PROTECTING MOBILE PRIVACY: YOUR SMARTPHONES, TABLETS, CELL PHONES AND YOUR PRIVACY"*

Chairman Franken, Ranking Member Coburn, and the distinguished members of the Subcommittee, thank you for the opportunity to testify about mobile privacy and the state of location tracking.

My name is Ashkan Soltani. I am a technology researcher and consultant specializing in consumer privacy and security. I have more than 15 years of experience as a technical consultant to Internet companies and federal government agencies. I received my masters degree in Information Science from the University of California at Berkeley, where I conducted extensive research and published two major reports on the methods by which users are tracked online and to what extent. Last year, I served as a staff technologist in the Division of Privacy and Identity Protection at the Federal Trade Commission on investigations related to Internet technology and consumer privacy. I have also worked as the primary technical consultant on The Wall Street Journal's *What They Know* series investigating issues relating to privacy

* This is an edited, reformatted and augmented version of a testimony presented May 10, 2011 before the Senate Judiciary Committee, Subcommittee on Privacy, Technology and the Law.

online. Recent revelations about how mobile devices handle sensitive data—particularly location information—have surprised consumers. Their devices often play a large role in their everyday activities, and many consumers show significant concern about who has access to their information.[2] Whether consumers understand these privacy risks and whether they have meaningful control over information access are critical questions for this Subcommittee.

I have been invited to testify about the current state of mobile privacy and location tracking from a technical perspective. First, I will describe location-based services and how a mobile device can determine its location. Second, I will discuss three recent issues that demonstrate how location data and other personal information are collected and shared in the current mobile ecosystem. Finally, I will discuss three broad implications for consumer mobile privacy and provide some suggestions for improvement.

A. Mobile Devices And Location-Based Services

Mobile devices today are powerful computing machines. Like desktop computers, many mobile devices run complex operating system platforms that allow third-party developers to create software applications to perform specialized tasks. Two of the most widely used mobile platforms, Apple iOS and Google Android, offer consumers hundreds of thousands of innovative applications to download and install onto their devices through the Apple App Store and Android Market. These include e-mail capabilities and productivity tools, mapping and navigation services, social media applications and games. However, unlike desktop computers, mobile devices are uniquely *mobile* which introduces unique privacy implications for their owners.

Consumers take their mobile phones and tablet computers with them nearly everywhere they go.[3] They often carry these devices in their pockets from their homes to their offices, while traveling by car or train, when on their way to daycare and to the grocery store. Mobile phones, in particular, are personal "always-on" devices; therefore, the location of these devices often closely mirrors that of their owners' locations and activities. The location of a mobile device at any given moment may not be particularly sensitive; However, the historical trail of past locations can reveal much about its user's behavior. In some cases, a person who has access to historical location data can infer trends that uniquely identify an individual. For example, if a mobile device's location is the same each work day, then consistently at another location every evening, it might expose the location of the device owner's workplace and home, respectively. An individual or organization with access to this information could then correlate it with public databases that could then be linked to a particular individual.[4] However, location-based services (LBS) are a major selling point for many mobile devices. These features quickly enable the discovery of nearby stores and restaurants, sharing of current location with friends and family by using "check-in" functionality within social networking applications, and easy directional navigation to desired destinations. In order to provide this functionality, the application or service provider needs to pinpoint and use the mobile device's location.

> *Note:* The icons in the margins below refer to the diagram in Appendix A and are used to direct attention to specific portions of the "Location Ecosystem."

There are four primary ways the location of a mobile device can be determined, depending on both its hardware and software capabilities.

1. *Global positioning system* (GPS) is a technology that allows a device to determine its location by triangulating GPS satellite signals, which are typically accurate to within a few meters. While nearly all smartphones manufactured today contain a built-in GPS chip, many mobile devices (*e.g.*, laptops) typically do not. While GPS allows for high accuracy of location, it is often unavailable indoors and its high consumption of battery life often compels users to turn off GPS until they require it.

2. *Wireless carriers* can help mobile devices determine location by using information about the signals of nearby cell phone towers. This is called cellular geolocation. Cellular phone towers act as known "landmarks" since they have fixed locations. This property enables wireless carriers to triangulate a device's location anytime the device is powered on. Mobile phones can send a query to the carrier to request the physical coordinates of towers within range and then calculate its position as best as possible. This technique is generally less accurate than GPS and varies widely depending on the density of cell towers in a given area.

3. *Location providers* are services that allow devices to determine location via a variety of methods, which include cellular, Wi-Fi and Internet Protocol (IP) based methods.

Companies such as Google, Apple, and Skyhook can act as location providers by compiling extensive databases that correlate Wi-Fi access points and cell phone towers with their physical locations. Mobile devices then query these databases with information about nearby "wireless landmarks" (*i.e.,* Wi-Fi access points and cell phone towers) in order to obtain their current location. As a result, the location provider is able to infer the current location of the mobile device as well as enhance its own location database with any additional "wireless landmarks" provided with the query.

4. *Location aggregators* are a separate class of location service providers that obtain location information via direct arrangements with wireless carriers. As such, device location is obtained directly from triangulation of nearby cellular tower data and does not rely on the handset to be "aware" of its present location. This enables features such as 'geofencing,' which is the ability to notify a third party whenever a device enters geographic area without requiring a specialized application on the phone. Location aggregators occupy a unique niche in this marketplace as they have a detailed "carrier view" vantage point across all of their participating partners, and they provide data to third party applications and websites directly.

B. How Mobile Device Location Is Collected and Shared

1. By Location Providers

The process by which Location Providers gather data raises significant privacy concerns. Much of the initial public concern focused on Google's reported collection of consumer information when it mapped wireless landmarks like cell towers and Wi-Fi access points by using employee-driven automobiles that were equipped with special sensors.[5] More recently, location providers began distributing the work by using their customers' mobile devices as "scouts in the field" in order to compile their databases of the physical locations of wireless landmarks. This "crowdsourcing" of location data has introduced additional privacy concerns. By leveraging consumers' mobile devices as scouts, location providers consequently receive the location of the mobile device as they report their findings.[6] Consumers have the option to "opt-out" of this practice; however, background collection and transmission of location information is enabled by default for most location providers.[7]

Even the notice that is offered may also be inadequate for meaningful choice. Figure 1 below compares the Google Android platform's permission screen informing users of the background collection of location data to the comparable screen on the Apple iOS platform. A customer would have to read Apple's lengthy software license agreement to learn that disabling location services means disabling the background collection of location data.

In addition, a mobile device user's attempt to "opt-out" may be ineffective. In April 2011, The Wall Street Journal reported that Apple iPhone devices would still collect and transmitting this information, even when users' had affirmatively set the location services to "off." That is, even when consumers elected to disable collection of their device location, their iPhones had continued to record and transmit location services information to Apple's servers.[8] Surprisingly, this scenario conflicts with a July 12, 2010 letter from Apple's General Counsel to Representatives Ed Markey and Joe Barton which stated that "Apple automatically collects this information only (1) if the device's location-based services capabilities are toggled to 'On' and (2) the customer uses an application requiring location-based information."[9]

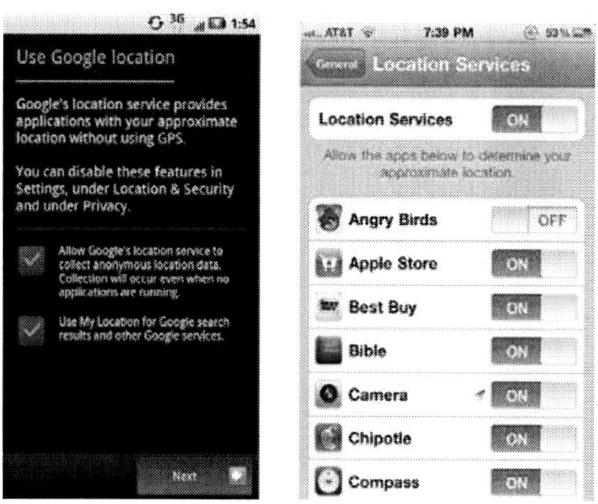

Figure 1. Permission screens controlling location service on the Android and iPhone platforms. (Location Services and subsequent collection is ON by default on both platforms).

2. In Local Cache Files on the Device

In order to improve the speed of location look-ups and to further reduce battery consumption, many mobile platform developers design their systems to keep a local copy - a "cache" - of location information from previous queries on the mobile device. This allows a mobile device to determine its location without having to re-query the location provider every time it's near a previously seen landmark. Like any repository of sensitive information, this cache of location data poses potential privacy issues. As mentioned previously, a person who is able to gain access to this database might be able to determine the user's past whereabouts (subject to the historical length of the cache). In addition, last month, researchers identified a cache of location data that includes a full year's worth of location history stored on their Apple iPhone device.[10] This data had been recorded by iPhones even when a user elected to disable location services. This effectively means that, in addition to there being no meaningful mechanism by which consumers can disable the background collection of location data by location providers, they also lack a meaningful mechanism to disable the collection of location data in a cache file. The researchers also found a copy of this same cache file stored insecurely on computers that had been used to synchronize or backup their iPhones, iPads, and other iOS devices.[11] By analyzing the data stored in this cache, which is a record of nearby cellular towers and Wi-Fi access points the phone encountered, the researchers were able to re-create a map of their previous travels from Washington DC to New York, as shown below in Figure 2. They also publicly released a tool that consumers could use to easily access and visualize their own location histories.[12]

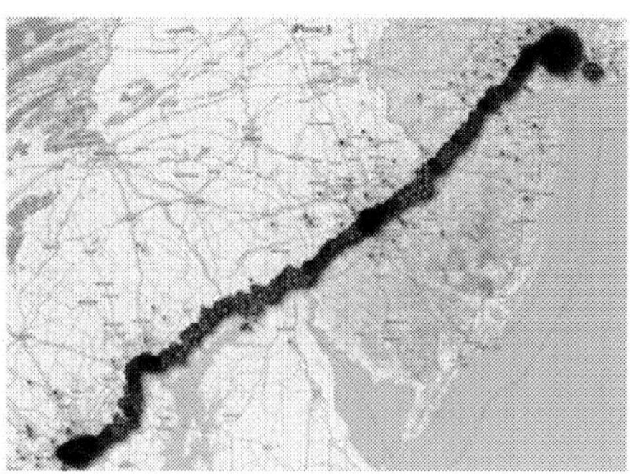

Figure 2. Map of researchers' whereabouts, inferred from local iPhone cache.[13]

Further research into competing platforms showed that Apple was not alone in this practice. Google and Microsoft smartphones also cache location histories, although the retention period for this information on these platforms appears to be shorter.[14]

It's worth noting here that while the recent "discovery" of local location caches has better informed the public about the issue, researchers and law enforcement have been aware of this practice for some time.[15] In addition to location history, researchers have repeatedly demonstrated that personal information such as email, text messages, browsing history, photos, and passwords can be recovered easily with physical access to the devices and, in some cases, remotely.[16] Surprisingly, this is even true for applications typically thought to be impervious to monitoring, such as the encrypted voice calling program Skype.[17]

3. By Smartphone Applications

In addition to storing location data locally and transmitting it to Location Providers, many users' smartphones will transmit their location and other sensitive data to numerous third parties via the use of third-party applications, such as games and other software programs. The specific parties and amount of information will vary depending on the specific "apps" used. However, the practice of transmitting potentially sensitive data off of the device is common for most applications.

In a survey of the 101 popular iPhone and Android phone apps in December 2010, The Wall Street Journal found that 47 of them transmitted the phone's location and 56 also transmitted identifiers (such as hardware serial numbers) to third parties.[18] Sometimes this information would go to the application developer's server, such as Yelp.com when using the Yelp "app." Other times, the location would be shared by the app further afield to its advertising partners without clear indication to the end-user. Forty-five apps had no discernible privacy policies, and neither Apple nor Google requires apps to have privacy policies.

While user consent is typically required before applications are allowed to access location information, the purpose may not always be apparent to the user, and the user may have no indication that this information will subsequently be disclosed to third parties. For example, one iPhone app called Ninjump—a game—accesses and sends the a mobile device's location information to its mobile ad provider.[19] Most users would probably be befuddled about why an action game would ever need to access their location or disclose it to others, even if they consented to the initial collection of this information. Data sharing isn't limited to location information. Applications can access and transmit data which includes text messages, emails, phone numbers, contacts stored, and even browser history stored on the device, as well as any information users knowingly enter in the process of using the app.[20] Some of this sharing may be expected, while other times it may be surprising. One example is where a popular social networking application had uploaded entire copies of users' address books to Facebook's servers.[21]

C. Implications for Consumer Privacy

These recent issues demonstrate key points of contention between consumers privacy and business interests.

1. Existing Notice and Choice Mechanisms Are Insufficient

Mobile apps and platforms do not provide consumers with sufficiently detailed notices about how their location and other sensitive information will be collected and used. Notice requirements vary from platform-to-platform. However, many disclosures related to privacy, such as data retention and sharing, frequently go unmentioned. The notices also rarely differentiate between first and third party data uses nor do they reveal business partners, like ad networks, by name. As such, consumers are unable to make meaningful choices regarding their privacy risks when using mobile devices.

For example, with the exception of real-time location data, Apple's iOS platform for iPhones does not disclose to users what other location information may be accessed and shared by applications upon download. The iOS platform also does not inform users if an app will collect information from their address books, calendars, or other data from their iPhone.Consumers are given a chance to "click through" to discover individual app privacy policies, but these are often long legal statements that are particularly difficult to read on a small mobile screen,[22] when they're even available. Comparatively, the Android platform allows more descriptive notices informing users of the data an app will collect. Although many of the terms used in these notice are still very technical in nature and can appear cryptic for a lay user to understand. While mobile platforms today allow users to first review these disclosure notices before they install an app. But they also all adopt a "take it or leave it" approach to application permissions: the user can either allow access to all of the information the app requests, or deny all access (and thus not install the app). Granular permissions are not typically made available. That is, users are forced to give up their location information if they want to play the Ninjump game.

2. Collected Location Information Can Be Sensitive

Some industry players dismiss the recent concern about location privacy by saying that the information collected is not actually *device* location information. In Apple's Q&A on location data, they say that some of the collected information is about network equipment "some of which may be located more than one hundred miles away."[23]

While this may be true for cellular location in sparse rural areas, many urban environments yield device location measurements as accurate as 50 to 200 feet.[24] Since Wi-Fi is a short-range communication, knowing even one nearby Wi-Fi signal can typically pin the user within 100 feet.

As a quick demonstration, I recorded my device's location while sitting on a bench in the lobby of Hart Senate Office Building. Using GPS, my location was accurately reported to within 20 meters, as indicated by the small circle at the center of the left image in Figure 3 below. The right image shows nearly the exact same location found using Wi-Fi geolocation, which only uses a location database maintained by Google.

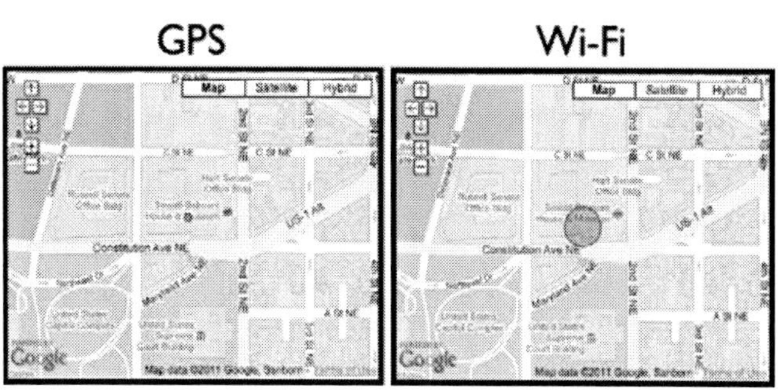

Figure 3. Comparing the accuracy of GPS and Wi-Fi based geolocation techniques.[25]

Quite a lot of information can be deduced from trails of historical location data. People are creatures of habit,[26] and it would often be easy to deduce where an individual works from her location on weekdays from 9am-5pm or, from the same nightly location, where she sleeps.

These two pieces of information start to form a picture of who the device owner is.

3. Location Data Can Be Tied to Consumer Identities

Industry also argues that location data cannot be associated with consumers' real identities, and that this data if often simply "anonymous usage statistics."[27]

However, to the degree that this data is also associated with unique identifiers—such as serial numbers or IP addresses that can later be linked back to an individual device or person[28]—it becomes difficult to refer to it as "anonymous information."[29]

Identifiers enable further correlations with additional information generated via other channels, such as subscriber information (from a wireless carrier), login credentials (from phones that sync their e-mail or calendars), or even in some cases name, credit card or address information used in the app marketplace.

For example, research recently demonstrated that "anonymous" device identifiers can easily be correlated to user's location and identity" in the form of pseudonyms and Facebook profiles with a reasonable degree of likelihood.[30]

Whether re-identification is possible depends on what other information is available, which itself hinges on the data retention and security practices of multiple participants in this ecosystem.

It is rarely the case that information should be called "anonymous," since there is nearly always some small chance of re-identification.

Fortunately, at least some in industry share this view. When asked about the anonymity of location, the CEO of Location Provider Skyhook Jay Yarao stated:

> "I[f] you associate any history of a user at all it's very easy to, after the fact, figure out the name of that user. So when you hear companies like Microsoft and Google say, 'We're anonymizing the data,' it doesn't matter.

If there's a location history, all I do is look at past 9 o'clock and there's a 95% chance that you went home. And I will look at that, and I will look up that address and I will know who you are. And as you start adding more and more data, I match that with where you work and now I know this is you."[31]

Conclusion

As mobile devices become more powerful—and more ingrained in the way consumers work and play—information about where a device is located becomes an ever more valuable input for commercial activity. But at the same time, consumers have expressed significant concern about how their devices expose sensitive information about them in ways they might not expect. Consumers need to be able to trust their devices in order to take full advantage of all the benefits mobile technology has to offer. To better protect consumer privacy going forward, I offer four suggestions:

1) Mobile platform providers and application developers should work together to provide consumers with more transparency into exactly what data are collected, how they are stored, to whom they are transmitted, and how they are secured and used.
2) Certain disclosures should be mandatory, such as clearly differentiating between first and third party uses of all potentially sensitive data, and also between active use and passive background activity.
   Precise definitions for "location" and "identity" should be provided.
3) Providers and developers should also work to ensure that the information consumers entrust with them are handled securely and in line with their expectations.
4) Providers and developers should also offer meaningful choice, such as granular permissions and working opt-outs, to consumers so they can make effective, privacy-conscious decisions in the marketplace.

Thank you for the opportunity to testify here today. Mobile privacy is a very nuanced issue, even for us technologists, so I thank the subcommittee for their attention on this increasingly important problem. I will be happy to answer any further questions.

Appendix A: Flow of Location Data in Mobile Ecosystem

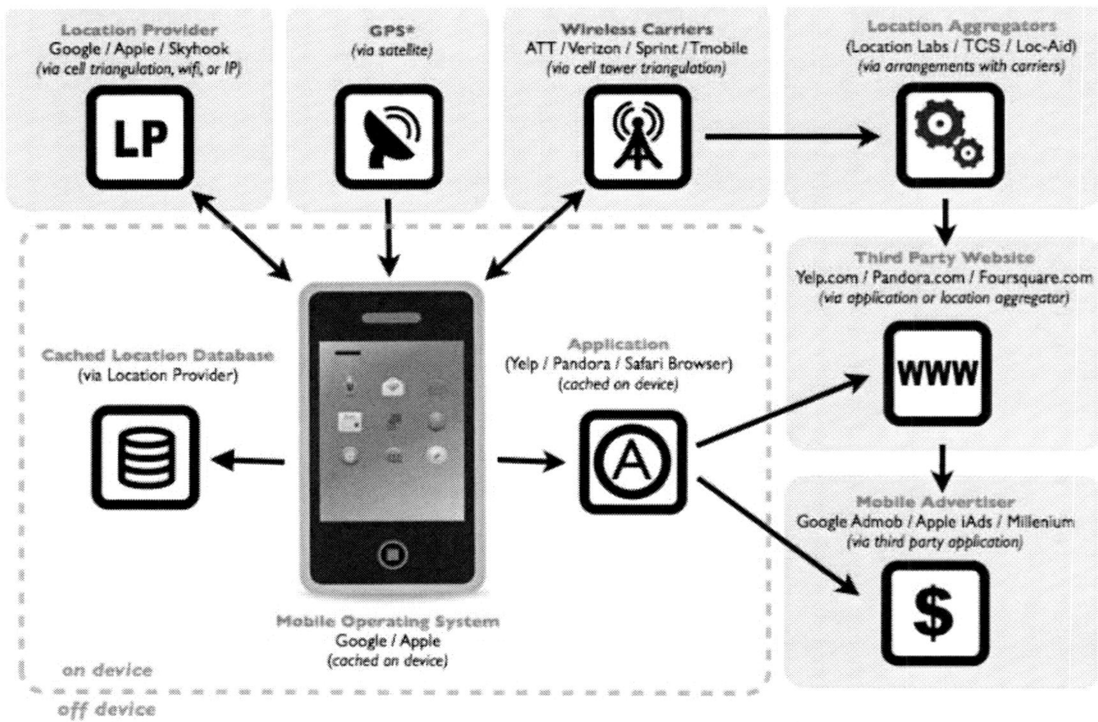

Note: GPS satellites provide a location signal, but do not have access to device location.

End Notes

[1] My oral and written testimony here today to the Subcommittee represents my own personal views, and does not reflect the views of any of the organizations that I have consulted or worked for in the past.

[2] Tsai, Janice Y., Kelley, Patrick Gage, Cranor, Lorrie Faith, Sadeh, Norman. Location Sharing Technologies: Privacy Risks and Controls. (2009). From http://repository.cmu.edu/isr/85/

[3] Three in five mobile phone owners say they carry their phones at all times, even inside the home. See: Stanton, D. (2008, September 8). New Study Shows Mobile Phones Merging New, Established Roles. Knowledge Networks. From http://www.knowledgenetworks.com/news/ releases/2008/091808_mobilephones.html

[4] Golle, Philippe and Kurt Partridge. On the Anonymity of Home/WorkLocation Pairs. From http://xenon.stanford.edu/~pgolle/papers/commute.pdf (Researchers demonstrate it may be possible to associate home/work location pairs to individuals' identity.)

[5] Stone, Brad. (2010, May 14). Google Says It Collected Private Data By Mistake. From http://www.nytimes.com/2010/05/15/business/15google.html

[6] Valentino-Devries, Jennifer. (2011, April 23). Google Defends Way It Gets Phone Data. From http://online.wsj.com/article/SB10001424052748703387904576279451001593760.html

[7] Google's default is enabled by means of a pre-selected check box during the initial product setup which a user has to actively 'uncheck'. See Figure 1. The FTC has raised concerns about "pre-checked" dialogues as a mechanism for affirmative consent in a recent settlement with Google and their Buzz social networking product. See http://www.ftc.gov/os/caselist/1023136/ 110330googlebuzzcmpt.pdf at page 4.

[8] Valentino-Devries, Jennifer. (2011, April 25). IPhone Stored Location in Test Even if Disabled. From http://online.wsj.com/article/SB10001424052748704123204576283580249161342.html and Apple. (2011, April 27). Apple Q&A On Location Data. From http://www.apple.com/pr/library/2011/04/27location_qa.html.

[9] Apple Inc's Response For Information Regarding Its Privacy Policy and Location-Based Services. (2010, July 12). From http://markey.house.gov/docs/applemarkeybarton7-12-10.pdf at page 7.

[10] Allan, Alasdair and Pete Warden. (2011, April 20). Got An iPhone or 3G iPad? Apple Is Recording Your Moves. From http://radar.oreilly.com/2011/04/apple-location-tracking.html.

[11] Apple announced a fix for this bug which reduces the size of the location database cache, stops transfer to iTunes when you connect your device to a computer, and deletes the cache entirely when you turn Location Services off. However, this fix doesn't apply to older 2G and 3G devices. Chen, Jacqui. (2011, May 05). iOS 4.3.3 is out with location tracking fixes for iPhone, iPad. From http://arstechnica.com/apple/news/2011/05/ios-433-is-out-with-location-trackingfixes-for-iphone-ipad.ars

[12] Warden, Pete. (2011, April 20). iPhoneTracker. From http://petewarden.github.com/iPhoneTracker/

[13] Allan, Alasdair. (2011, April 20). Got an iPhone or 3G iPad? Apple Is Recording Your Moves. From http://radar.oreilly.com/2011/04/apple-location-tracking.html

[14] Gohring, Nancy. (2011, April 29). Microsoft Admits To More Windows Phone Update Problems. From http://www.pcworld.com/article/226733/microsoft_admits_to_more_windows_phone_update_pro blems.html and Foresman, Chris. Android Phone Keeps Location Cache Too, But It's Harder To Access. From http://arstechnica.com/gadgets/news/2011/04/android-phones-keep-locationcache-too-but-its-harder-to-access.ars

[15] Levinson presented his research on the iPhone cache file at a conference six months ago and subsequently published his findings in December 2010. Levinson, Alex. (2011, April 21). Three Major Issues with the Latest iPhone Tracking "Discovery." From https://alexlevinson.wordpress.com/2011/04/21/3-major-issues-with-the-latest-iphone-trackingdiscovery/. Johnson, Bobbie. (2011, April 21). Researcher: iPhone Location Data Already Used By Cops. From http://gigaom.com/2011/04/21/researcher-iphone-location-data-already-used-bycops/.

[16] Edwards, Sarah. Inside the App: All Your Data are Belong to Me. From http://www.shmoocon.org/speakers#insideapp.

[17] A design vulnerability in the secure calling software Skype allows access to "full name, date of birth, city/state/country, home phone, office phone, cell phone and email addresses" of users because files on the device had insecure permissions and we stored in an unencrypted format. Case, Justin. (2011, April 15). (Updated) Exclusive Vulnerability In Skype For Android Is Exposing Your Name, Phone Number, Chat Logs, And A Lot More. From http://www.androidpolice.com/2011/04/14/exclusive-vulnerability-in-skype-for-android-is-exposing-your-name-phone-number-chat-logs-and-a-lot-more/

[18] Thurm, Scott and Yutari Iwatani Kane. (2010, December 17). Your Apps Are Watching You. From http://online.wsj.com/article/SB10001424052748704694004576020083703574602.html

[19] WSJ Blogs. (2011, December 17). What They Know Mobile. Ninjump. From http://blogs.wsj.com/wtk-mobile/2010/12/17/ninjump/

[20] Seriot, Nicolas. (2010). iPhone Privacy. From http://seriot.ch/resources/talks_papers/ iPhone Privacy.pdf

[21] Moos, Kurt von. (2010, February 26). Privacy Fails: How Facebook Steals Your Friends Numbers. From http://kurtvonmoos.com/facebook_steals_contact_info/

[22] This matter became the underlying premise of a popular television show parodying "Apple's ridiculous 55-page iTunes terms and conditions." O'Grady. Jason. D (2011, April 28) South Park parodies iTunes terms and conditions. From http://www.zdnet.com/blog/apple/south-parkparodies-itunes-terms-and-conditions/10043.

[23] Apple. (2011, April 27). Apple Q&A On Location Data. From http://www.apple.com/pr/library/2011/04/27location_qa.html

[24] Steve Lee, product manager for Google Maps for Mobile and Google Latitude said in a May 2010 email that Google had 300 million Wi-Fi networks in its database which could pinpoint a device's location to within about 100 feet. Efrati, Amir. (2011, May 1). Google Calls Location Data 'Valuable.' From http://online.wsj.com/article/SB10001 424052748703 703304576297450030517830.html

[25] The strongest Wi-Fi signal my device could detect was one of the "Odyssey" access points. Google's geolocation database reported the location of this access point (and thus my location) to within 36.0 meters as indicated by the circle in the right image (as identified by Google). 2693% of people return to the same locations: Song, C., Qu, Z., Blumm, N., and Barabási, A.L. Limits of predictability in human mobility. Science. 2010 Feb 19, 327(5968): 1018-21. From http://www.ncbi.nlm.nih.gov/pubmed/20167789

[27] "Google spokesman said it collects information anonymously." Kane, Yukari Iwatani, and Jennifer Valentino-DeVries. (2011, April 28). Jobs Tries to Calm iPhone Imbroglio. From http://online.wsj.com/article/SB10001424052748703367004576288790268529716.html

[28] While IP addresses can be dynamic, they can persist for days. IP addresses assigned to phones on the Verizon and Sprint do not change over a 2-day test period. See Balakrishnan, Mahesh, Iqbal Mohomed, and Venugopalan Ramusubramanian. (2009). Where's That

Phone? Geolocating IP Addresses on 3G Networks. From http://research.microsoft.com/en-us/um/people/maheshba/papers/ephemera-imc09.pdf

[29] The Dutch Data Protection Authority argues that MAC addresses, in combination with the ability to identify the location of wireless hardware, may by itself qualify as personal information. Preuschat, Archibald. (2011, April 20). Google Faces New Demands In Netherlands Over Street View Data. From http://online.wsj.com/article/0,, SB1 0001424052748703922504576273151673266520,00.html

[30] Recently, a researcher demonstrated that device IDs can be linked to GPS location (30%), Weak Identities (20%), and Facebook profiles (10%) using public game service OpenFeint. See Cortesi, Aldo. (2011, May 4). De-Anonymizing Apple UDIDs with OpenFeint. From http://corte.si/posts/security/openfeint-udid-deanonymization/index.html

[31] Yarao, Jay. (2011, April 28). Everything You Need To Know About How Phones Are Stalking You Everywhere. From http://www.businessinsider.com/skyhook-ceo-2011-4

In: Mobile Devices ISBN: 978-1-62618-304-9
Editors: A. R. Walker, D. Brown Miller © 2013 Nova Science Publishers, Inc.

Chapter 5

TESTIMONY OF ALAN DAVIDSON, DIRECTOR OF PUBLIC POLICY, GOOGLE. HEARING ON "PROTECTING MOBILE PRIVACY: YOUR SMARTPHONES, TABLETS, CELL PHONES AND YOUR PRIVACY"*

Chairman Leahy, Chairman Franken, Ranking Member Coburn, and members of the Committee:

I am pleased to appear before you this morning to discuss mobile services, online privacy, and the ways that Google protects our users' personal information. My name is Alan Davidson, and I am a Google's Director of Public Policy for the Americas. In that capacity, I oversee our public policy operations in the United States, and work closely with our legal, product, and engineering teams to develop and communicate our approach to privacy and security, as well as other issues important to Google and our users.

Google is most well known for our search engine, which is available to Internet users throughout the world. We also make Android, an open operating system for mobile devices that in a few short years has grown from powering one device (introduced in the fall of 2008) to over 170 devices today, created by 27 manufacturers. We also offer dozens of other popular services, from YouTube to Gmail to Google Earth. Our products are generally offered for free for personal use, and one supported by revenue from advertising and sales

* This is an edited, reformatted and augmented version of a testimony presented May 10, 2011 before the Senate Judiciary Committee, Subcommittee on Privacy, Technology and the Law.

to businesses. Protecting privacy and security is essential for Internet commerce. Without the trust of our users, we simply would not be able to offer these services or platforms because on the Internet, competing services are only one click away. If we fail to offer clear, usable privacy controls, transparency in our privacy practices, and strong security, our users will simply switch to another provider. This is as true for our services that are available on mobile devices as it is for those that are available on desktop computers. For this reason, location sharing on Android devices is strictly opt-in for our users, with clear notice and control.

In my testimony today, I'll focus on three main points:

- Location-based services provide tremendous value to consumers;
- Google is committed to the highest standards of privacy protection in location-based services; and
- Congress has an important role in helping companies build trust and create appropriate government access standards.

I. Location Based Services Provide Tremendous Value to Consumers

Mobile services are creating enormous economic benefits for our society. A recent market report predicts that the mobile applications market will be worth $25 billion by 2015. At Google, we have seen an explosion in demand for location-based services. People can use our services to find driving directions from their current location, identify a traffic jam and find an alternate route, and find the next movie time at a nearby theater. Location can even make search results more relevant: If a user searches for "coffee" from a mobile phone, she is more likely to be looking for a nearby café than for the website of a national coffee chain or the Wikipedia entry describing coffee's history. In the last year, a full 40% of Google Maps usage was from mobile devices. There are now 150 million active monthly Google Maps for Mobile users on Android, iPhone, BlackBerry, and other mobile platforms in more than 100 countries. Many third party applications also use location services to provide helpful products. For example, the U.S. Postal Service offers an application to help users find nearby post offices and collection boxes, based on their location. And if you want a Five Guys burger, their application will find a location for you, and even lets you order and pay in advance. Twitter allows users to "geotag" their tweets from their application, which can give followers important context and perspective. On smartphones like iPhone, Palm, and Android devices, services such as Yelp and Urbanspoon use location to provide relevant local search results, while applications like Foursquare let users find nearby friends. Mobile location data can even save lives. In the past, a parent's best hope of finding a missing child might have been a picture on a milk carton, but mobile location services may be changing that. Google works with the National Center for Missing and Exploited Children (NCMEC) in an ongoing partnership to develop technology solutions that help them achieve their mission. Today, modern tools and information can make NCMEC's AMBER alerts more effective and efficient by sending the alert to all users within one mile of an incident within seconds of the report through location-based targeting. Over time, the radius could be expanded, with speed and acceleration of distribution based directly on information received. Existing emergency notifications like AMBER alerts can be improved using location data. In crisis situations, people are increasingly turning to the Internet on mobile or desktop devices to find information. Within a few hours of the Japan earthquake, for example, we saw a massive spike in search queries originating from Hawaii related to "tsunami." We placed a location-based alert on the Google homepage for tsunami alerts in the Pacific and ran similar promotions across News, Maps, and other services. In cases like the Japanese tsunami or the recent tornadoes in the U.S., a targeted mobile alert from a provider like Google or from a public enhanced 911 service may help increase citizens' chances of getting out of harm's way. None of these services or public safety tools would be possible without the location information that our users share with us and other providers, and without the mobile platforms for businesses and governments to effectively reach the appropriate audience.

II. Google Is Committed to The Highest Standards of Privacy Protection in Location-Based Services

Google would not be able to offer these services or platforms or help create the economic and social value generated from location data if we lost the trust of our users. Thus, at Google, privacy is something we think about every day across every level of our company. It is both good for our users and critical for our business.

Privacy at Google begins with five core principles, which are located and available to the public at www.google.com/corporate/privacyprinciples.html:

- Use information to provide our users with valuable products and services.
- Develop products that reflect strong privacy standards and practices.
- Make the collection and use of personal information transparent.
- Give users meaningful choices to protect their privacy.
- Be a responsible steward of the information we hold.

As with every aspect of our products, we follow the axiom of "focus on the user and all else will follow." We are committed to using information only where we can provide value to our users. That's what we mean by our first principle.

For example, *we never sell our users' personally identifiable information.* This is simply not our business model.

To further guide us, under the second principle, we aim to build privacy and security into our products and practices from the ground up. From the design phase through launch, we consider a product's impact on our users' privacy. And we don't stop at launch; we continue to innovate and iterate as we learn more from users.

Our last three principles give substance to what we mean by privacy: We commit to *transparency*, *user control,* and *security.*

Internal Process and Controls

We also reflect these principles in our development process and employee training. As consumers become more reliant on services provided by third parties, consumer privacy relies increasingly on those parties' internal practices, process, and controls. As we recently explained, we have begun to implement even stronger privacy controls with a focus on people, training, and compliance.

We have developed a review process where all engineering projects leads are required to submit and maintain a Privacy Design Document detailing how their projects handle user data. These documents are reviewed by cross-functional working groups that can request code reviews and make recommendations to the product teams. Completion of Privacy Design Documents will also be reviewed by managers and an independent internal audit team. We have also enhanced our core training for engineers and others to create a greater focus on responsible collection, use, and handling of data.

All this process is aimed at ensuring that products match our philosophy and avoid mistakes that fracture user trust — like the launch of Google Buzz — which fall short of our standards for transparency and user control. To help make sure we live up to this promise, we entered into a consent decree with the Federal Trade Commission this year, under which we'll receive an independent review of the privacy procedures we have outlined above once every two years. In addition, we'll ask users to give us affirmative consent before we change how we share their personal information.

How our products reflect our principles — Opt-in controls on Android

Moving to our specific products, I'll focus first on an important area in which we are putting our principles to work, and where we are innovating on the broader privacy issues faced in the online world: Simple, opt-in controls for collection and use of location information on Android.

While location-based services are already showing great value to users, Google recognizes the particular privacy concerns that come with the collection and storage of location information. That's why we don't collect any location information — any at all — through our location services on Android devices unless the user specifically chooses to share this information with Google. We also give users clear notice and control; the set-up process asks users if they would like to "allow Google's location service to collect anonymous location data."

And even after opting in, we give users a way to easily turn off location sharing with Google at any time they wish. The location services in our Android operating system embody the transparency and control principles that we use to guide our privacy process.

Google is also very careful about how we use and store the data that is generated by location-based services. The location information sent to Google servers when users opt in to location services on Android is anonymized and stored in the aggregate and is not tied or traceable to a specific user. The collected information is stored with a hashed version of an anonymous token, which is deleted after approximately one week. A small amount of location information regarding nearby Wi-Fi access points and cell towers is kept on the Android device to help the user continue to enjoy the service when no server connection is available and to improve speed and battery life. This information on the device is likewise not tied or traceable to a specific user.

Global Positioning System (GPS) enabled devices can provide a highly accurate location using information from GPS satellites. But GPS can be slow and drain battery life and can take 10 seconds (and sometimes much longer) to "fix" a location. Furthermore, many devices are not GPS enabled or are used in situations where obtaining a GPS signal might not even be possible (*e.g.*, indoors, where there is no line of sight between the device and the satellites).

In order to serve devices that may not have GPS capabilities, or simply to avoid the delay and battery drain from GPS services, various companies have worked out alternatives to GPS. These are generally based around the idea of detecting nearby, publicly available signals from Wi-Fi access points and cell towers and using this data to quickly approximate a rough position, usually with less accuracy than GPS. By treating Wi-Fi access points or cell towers as beacons, devices are able to fix their general location quickly in a power-efficient way, even while they may be working on a more precise GPS-based location.

This can be done by using information that is publicly broadcast (for example, that list of Wi-Fi access points you see when you use the "join network" option on your computer). A database of known network locations is required to determine a user's estimated location from either Wi-Fi access point or cell tower information. Companies like Skyhook Wireless and Navizon compile such databases and license the data to many industry leaders.

Google has also created such location service called the Google Location Server — an Internet database on Google servers that uses Wi-Fi access points and cell towers to determine an estimated location and that uses GPS information to estimate road traffic. Device manufacturers can install the Google Network Location Provider application for Android (pursuant to a license with Google) on their devices.

This application can determine a user's estimated location using the Google Location Server, to make location information available to users whether they are indoors and outdoors, more quickly, and using less battery power than GPS services. This Network Location Provider is turned off by default, and can be turned on by the user during the phone's initial setup or in the device settings.

The Network Location Provider is off by default. The user can opt-in and turn on location services during the initial setup flow.

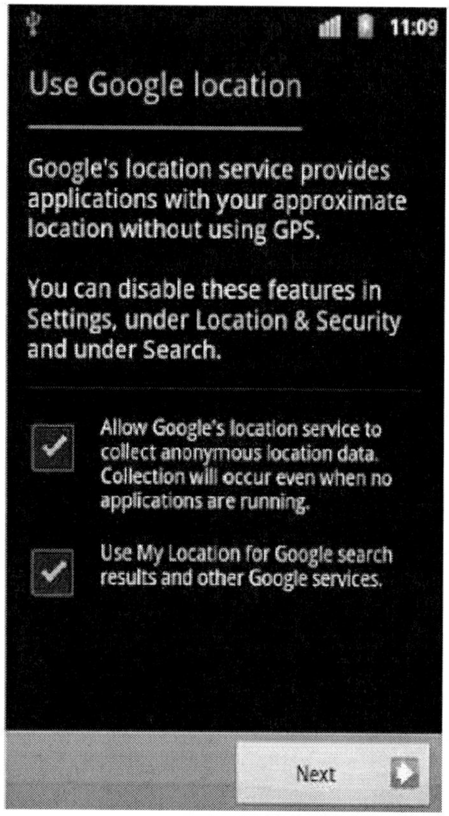
11:09
Use Google location
Google's location service provides applications with your approximate location without using GPS.
You can disable these features in Settings, under Location & Security and under Search.
Allow Google's location service to collect anonymous location data. Collection will occur even when no applications are running.
Use My Location for Google search results and other Google services.
Next

11:28
Location & security settings
My Location
Use wireless networks
Use GPS satellites
When locating, accurate to street level
Screen unlock
Change screen lock
Use visible pattern
Use tactile feedback
SIM card lock
Set up SIM card lock

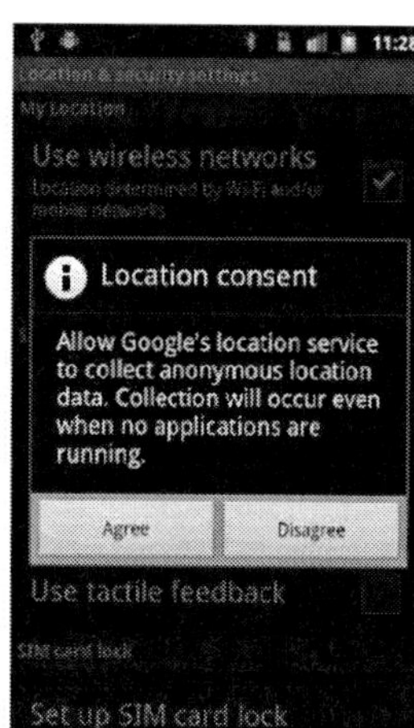
11:28
Location consent
Allow Google's location service to collect anonymous location data. Collection will occur even when no applications are running.
Agree
Disagree

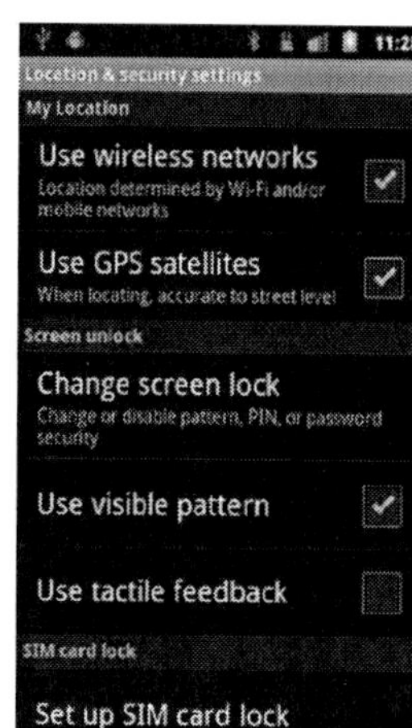
11:28
Location & security settings
My Location
Use wireless networks
Location determined by Wi-Fi and/or mobile networks
Use GPS satellites
When locating, accurate to street level
Screen unlock
Change screen lock
Change or disable pattern, PIN, or password security
Use visible pattern
Use tactile feedback
SIM card lock
Set up SIM card lock

The user can opt-in to turn on the Network Location Provider on their Android phone from within the device settings.

The Android operating system is built on the principle of openness, with the goal of encouraging developer innovation and a vibrant ecosystem for users. With this principle in mind, Google does not decide which applications can access location or other user information from the device. Instead, the Android operating system uses a permissions model in which the user is automatically informed of certain types of information an application will be able to access during the application installation process.

This permissions model is designed to empower users to make their own decision on whether or not to trust an application with the information requested.

The user may choose to trust the application by completing the installation or the user may choose to cancel the installation. An application can only access the device's GPS location or the device's network location if it displays a permission to the user at time of installation.

When Google creates an Android application, like the Google Maps for mobile application, Google is responsible for how the application collects and handles data and for the privacy disclosures made to users. Most Google-developed Android applications are subject to the Google Mobile Terms of Service and the Google Mobile Privacy Policy, unless Google has created a custom terms of service and privacy policy for the application. Google privacy policies are also clearly displayed to the user when the user first signs into the Android device.

When an Android application is not developed by Google, the application developer bears the responsibility for the design of the application, which includes responsibility for how the application collects and handles user data and the privacy disclosures made to users. If the user chooses to trust an application with location information by proceeding with the installation after viewing the location-related permissions, then that application could potentially store this location information on the device or transmit the information off the device if the application also has the Internet access permission.

Google does not control the behavior of third party applications or how they handle location information and other user information that the third party application obtains from the device, even though Google strongly encourages application developers to use best practices as described in this_Google blog post.

How Our Products Reflect Our Principles — Encryption and Two-Step Verification

Along with transparency and user control, strong security for users of Google's services to protect against hackers and data breach is vital. Nothing can erode trust faster than personal information falling into the hands of hackers. Google faces complex security challenges while providing services to millions of people every day, and we have world-class engineers working at Google to help secure information.

For example, Google is the first (and only) major webmail provider to offer session-wide secure socket layer (SSL) encryption *by default*. Usually recognized by a web address starting with "https" or by a "lock" icon, SSL encryption is regularly used for online banking or transactions. As our Gmail lead engineer wrote:

> In 2008, we rolled out the option to always use https — encrypting your mail as it travels between your web browser and our servers. Using https helps protect data from being snooped by third parties .... We initially left the choice of using it up to you because there's a downside: https can make your mail slower since encrypted data doesn't travel across the web as quickly as unencrypted data. Over the last few months, we've been researching the security/latency tradeoff and decided that turning https on for everyone was the right thing to do.

We hope other companies will soon join our lead.

We also hope to see our competitors adopt another security tool we offer our users: encryption for search queries. Users can simply type "https://encrypted.google" into their browsers to navigate to the version of Google Search that encrypts search queries and results. As we said in our blog post about encrypted search, "an encrypted connection is created between your browser and Google. This secured channel helps protect your search terms and your search results pages from being intercepted by a third party on your network."

And in March of last year Google introduced a system to notify users about suspicious activities associated with their accounts. By automatically matching a user's IP address to broad geographical locations, Google can help detect anomalous behavior, such as a log-in appearing to come from one continent only a few hours after the same account holder logged in from a different continent. Thus, someone whose Gmail account may have been compromised will be notified and given the opportunity to change her password, protecting her own account and her Gmail contacts.

Finally, we recently released 2-step verification for consumer Gmail accounts, which allows users who are concerned about the security of their account to use a password plus a unique code generated by a mobile phone to sign in. It's an extra step, but it's one that significantly improves the security of a Google Account. Now, if someone steals or guesses a Gmail user's password, the potential hijacker still cannot sign in to the user's account because the hijacker does not have the user's phone. We are already hearing stories from our users about how this extra layer of security has protected them from phishing attacks or unauthorized access.

III. Congress Should Act to Build Trust and Create Appropriate Government Access Standards

Congress has a vital role to play in encouraging responsible privacy and security practices, both by bringing attention to these issues and through appropriate legislation.

As a start, Google supports the development of comprehensive, baseline privacy framework that can ensure broad-based user trust and that will support continued innovation and serve the privacy interests of consumers. Some key considerations in this area include:

- *Even-handed application.* A pro-innovation privacy framework must apply even-handedly to all personal data regardless of source or means of collection. Thus, offline and online data collection and processing should, where reasonable, involve similar data protection obligations.
- *Recognition of benefits and costs.* As with any regulatory policy, it is appropriate to examine the benefits and costs of legislating in this

area, including explicit attention to actual harm to users and compliance costs.

- *Consistency across jurisdictions.* Generally, Internet users neither expect nor want different baseline privacy rules based on the local jurisdiction in which they or the provider reside. Moreover, in many instances, strict compliance with differing state or national privacy protocols would actually diminish consumer privacy, since it would require Internet companies to know where consumers are located at any given time.

We also suggest two concrete areas where Congress can act immediately to strengthen Americans' privacy protections and provide consistency for providers:

We pride ourselves at Google for industry-leading security features, including the use of encryption for our search and Gmail services. But we need help from the government to help ensure that the bad acts of criminal hackers or inadequate security on the part of other companies does not undermine consumer trust for all services. Moreover, the patchwork of state law in this area leads to confusion and unnecessary cost. Congress should therefore promote uniform, reasonable security principles, including data breach notification procedures.

Finally, the Electronic Communications Privacy Act, the U.S. law governing government access to stored communications, is outdated and out of step with what is reasonably expected by those who use cloud computing services. ECPA worked well for many years, and much of it remains vibrant and relevant. In significant places, however, a large gap has grown between the technological assumptions made in ECPA and the reality of how the Internet works today, leaving us in some circumstances with complex and baffling rules that are both difficult to explain to users and difficult to apply.

As part of the Digital Due Process coalition, we are working to address this issue. The Digital Due Process coalition includes members ranging from AT&T to Google to Americans for Tax Reform to the ACLU. It has put forward common sense principles that are designed to update ECPA, while ensuring that government has the legal tools needed to enforce the laws. Particularly relevant to today's hearing, the coalition seeks to:

- Create a consistent process for data stored online. Treat private communications and documents stored online the same as if they were stored at home and require a uniform process before compelling a service provider to access and disclose the information.

- Create a consistent process for location information. Create a clear, strong process with heightened standards for government access to information regarding the location of an individual's mobile device.

Advances in technology rely not just on the smart engineers who create the new services, but also on smart laws that provide the critical legal underpinning for continued innovation and adoption of the

technology. We hope to work with this Committee and with Congress as a whole to strengthen these legal protections for individuals and businesses.

* * *

I look forward to answering any questions you might have about our efforts. And Google looks forward to working with members of the Committee and with Congress in the development of valuable online services and strong privacy and security protections for users.

Thank you.

In: Mobile Devices ISBN: 978-1-62618-304-9
Editors: A. R. Walker, D. Brown Miller © 2013 Nova Science Publishers, Inc.

Chapter 6

TESTIMONY OF DR. GUY "BUD" TRIBBLE, VICE PRESIDENT FOR SOFTWARE TECHNOLOGY, APPLE INC.. HEARING ON "PROTECTING MOBILE PRIVACY: YOUR SMARTPHONES, TABLETS, CELL PHONES AND YOUR PRIVACY"*

Good morning Chairman Franken, Ranking Member Coburn, and Members of the Subcommittee. My name is Bud Tribble, and I am Vice President for Software Technology for Apple Inc. On behalf of Apple, I thank you for the opportunity to address this important subject.

APPLE'S COMMITMENT TO PROTECTING OUR CUSTOMERS' PRIVACY

Apple is deeply committed to protecting the privacy of our customers who use Apple mobile devices, including iPhone, iPad and iPod touch. Apple has adopted a comprehensive privacy policy for all its products and implemented industry-leading privacy features in its products to protect our customers' personal data. We are also deeply committed to meeting our customers'

* This is an edited, reformatted and augmented version of a testimony presented May 10, 2011 before the Senate Judiciary Committee, Subcommittee on Privacy, Technology and the Law.

demands for prompt and accurate location-based services. These services offer many benefits to our customers by enhancing convenience and safety for shopping, travel and other activities.

To meet these goals, Apple provides easy-to-use tools that allow our consumers to control the collection and use of location data on all our mobile devices. We do not share personally identifiable information with third parties for their marketing purposes without consent, and we require third-party application developers to agree to specific restrictions protecting our customers' privacy. Apple is constantly innovating new technology, features and designs to provide our customers with greater privacy protection and the best possible user experience.

Apple welcomes inquiries about how it protects its customers' privacy while providing reliable and fast location-based services. For instance, Apple provided on July 12, 2010 to Representatives Barton and Markey a detailed description of its collection and use of location-based information. I testified regarding the same topic before the Committee on Commerce, Science, and Transportation on July 27, 2010. And on April 27, 2011, Apple released a public response to recent questions regarding the collection and use of location information. A copy of that response is attached to this testimony as Exhibit A. The initial point made in that response should be emphasized: Apple does not track users' locations – Apple has never done so and has no plans to ever do so.

In my testimony today, I would like to address the following topics:

1) Apple's Privacy Policy;
2) Apple's collection, storage and use of location information on Apple mobile devices; and
3) the use of location information by third-party applications and the iAd Advertising Network.

I. Apple's Privacy Policy

Apple has a single Customer Privacy Policy (the "Policy") that applies across all Apple businesses and products, including the iTunes Store and App Store. The Policy, written in easy-to-read language, details what information Apple collects and how Apple and its partners and licensees may use the information. The Policy is available from a link on every page of Apple's website.[1]

The Policy includes the following provision regarding location-based information:

> To provide location-based services on Apple products, Apple and our partners and licensees may collect, use, and share precise location data, including the real-time geographic location of your Apple computer or device. This location data is collected anonymously in a form that does not personally identify you and is used by Apple and our partners and licensees to provide and improve location-based products and services. For example, we may share geographic location with application providers when you opt in to their location services.
>
> Some location-based services offered by Apple, such as the MobileMe "Find My iPhone" feature, require your personal information for the feature to work.

This provision incorporates similar language regarding location-based information that appears in Apple End User Software License Agreements ("SLAs") for products that provide location-based services. For example, the current iPhone SLA states:

> Apple and its partners and licensees may provide certain services through your iPhone that rely upon location information. To provide and improve these services, where available, Apple and its partners and licensees may transmit, collect, maintain, process and use your location data, including the real-time geographic location of your iPhone, and location search queries. The location data and queries collected by Apple are collected in a form that does not personally identify you and may be used by Apple and its partners and licensees to provide and improve location-based products and services. By using any location-based services on your iPhone, you agree and consent to Apple's and its partners' and licensees' transmission, collection, maintenance, processing and use of your location data and queries to provide and improve such products and services. (Emphasis exists in the SLA.) You may withdraw this consent at any time by going to the Location Services setting on your iPhone and either turning off the global Location Services setting or turning off the individual location settings of each location-aware application on your iPhone. Not using these location features will not impact the non location-based functionality of your iPhone. When using third party applications or services on the iPhone that use or provide location data, you are subject to and should review such third party's

terms and privacy policy on use of location data by such third party applications or services.

The Policy includes the following provision regarding third-party products, such as iPhone apps:

> Apple websites, products, applications, and services may contain links to third-party websites, products, and services. Our products and services may also use or offer products or services from third parties - for example, a third-party iPhone app. Information collected by third parties, which may include such things as location data or contact details, is governed by their privacy practices. We encourage you to learn about the privacy practices of those third parties.

The Policy also includes the following language regarding mobile advertisements, such as those served through Apple's iAd service:

> Apple and its partners use cookies and other technologies in mobile advertising services to control the number of times you see a given ad, deliver ads that relate to your interests, and measure the effectiveness of ad campaigns. If you do not want to receive ads with this level of relevance on your mobile device, you can opt out by accessing the following link on your device: http://oo.apple.com. If you opt out, you will continue to receive the same number of mobile ads, but they may be less relevant because they will not be based on your interests. You may still see ads related to the content on a web page or in an application or based on other non-personal information. This opt-out applies only to Apple advertising services and does not affect interest-based advertising from other advertising networks.

The Policy identifies a dedicated page on Apple's website where customers may submit privacy-related inquiries and comments. Apple monitors these submissions and responds to appropriate inquiries in a timely manner. Customers may also address privacy concerns to TRUSTe, Apple's third-party privacy monitor. A link to TRUSTe is displayed within the Policy.

As noted above, customers may access the Policy from every page on Apple's website. The Policy also was placed where Apple believed the largest number of customers would see it: the iTunes Store.

Customers attempting to open a new iTunes Store account are directed to a webpage titled: "iTunes Store Terms & Conditions and Apple's Privacy Policy." They are asked to click the same unchecked agreement box stating: "I have read and agree to the iTunes Terms and Conditions and Apple's Privacy Policy."

Apple updated the Policy on June 21, 2010.[2] The first time each existing iTunes Store customer logged on to the iTunes Store after that date, the iTunes Store displayed a message that prompted the customer to review the iTunes Store Terms and Conditions. The message stated:

> iTunes Store Terms and Conditions have changed. Please read and agree to the terms and conditions below to continue using the iTunes Store.

Customers were asked to click an unchecked agreement box stating: "I have read and agree to the iTunes Terms and Conditions and Apple's Privacy Policy." Customers who do not agree to the Terms and Conditions and the Policy are not be able to use the iTunes Store (e.g., cannot make purchases on the iTunes Store or the App Store), but they may continue to use iTunes software.

II. Location Information and Location-Based Services for Mobile Devices

Apple began providing location-based services in January 2008. These services enable applications that allow customers to perform a wide variety of useful tasks such as getting directions to a particular address from their current location or finding nearby restaurants or stores.

Apple offers location-based services on a variety of mobile devices, including the iPhone 3G, iPhone 3GS, iPhone 4 CDMA and GSM models, iPad Wi-Fi + 3G, iPad 2 Wi-Fi and 3G and, to a more limited extent, older models of the iPhone, the iPad Wi-Fi, and iPod touch.

All of Apple's mobile devices run on Apple's proprietary mobile operating system, iOS. Apple released iOS 4.1 on September 8, 2010. Apple released the current versions, iOS 4.3.3 and 4.2.8 (for the iPhone 4 CDMA model), on May 4, 2011. Currently, iOS 4.3.3 may be run on iPhone 3GS, iPhone 4 GSM model, iPod touch 3rd and 4th generations, iPad, and iPad 2. My testimony focuses on iOS 4.1 and later versions, including the free iOS update Apple released on May 4, 2011.

A. Privacy Features

Apple has designed features that enable customers to exercise control over the use of location-based services.

First, Apple provides its customers with the ability to turn "Off" all location-based service capabilities with a single "On/Off" toggle switch. For mobile devices, the toggle switch is in the "Location Services" menu under "Settings." As described more fully below, when this toggle is switched "Off,"

1) iOS will not provide any location information to any applications, including applications that may have previously received consent to use location information;
2) iOS will not collect or geo-tag information about nearby Wi-Fi hotspots or cell towers; and
3) iOS will not upload any location information to Apple from the device.

Second, Apple requires express customer consent when any application requests location-based information for the first time. When an application requests the information, a dialog box appears stating: "[Application] would like to use your current location." The customer is asked: "Don't Allow" or "OK." If the customer clicks on "Don't Allow," iOS will not provide any location-based information to the application. This dialog box is mandatory—neither Apple's applications nor those of third parties are permitted to override the notification.

Third, iOS 4 permits customers to identify individual applications that may not access location-based information, even if Location Services is "On." The Location Services settings menu provides an "On/Off" toggle switch for each application that has requested location-based information. When the switch for a particular application is "Off," no location-based information will be provided to that application.

Fourth, Customers can change their individual application settings at any time. An arrow icon alerts iOS 4 users that an application is using or has recently used location-based information. This icon will appear real-time for currently running applications and next to the "On/Off" switch for any application that has used location-based information in the past twenty-four hours. Finally, customers can use Restrictions, also known as Parental Controls, on a mobile device to prevent access to specific features, including Location Services. When a customer enables Restrictions, the customer must enter a passcode (this passcode is separate from the device passcode that the customer may set).

If the customer turns Location Services off and selects "Don't Allow Changes," the user of the device cannot turn on Location Services without that passcode.

B. Location Information

1. Crowd-Sourced Database of Cell Tower Location and Wi-Fi Hotspot Information

Customers want and expect their mobile devices to be able to quickly and reliably determine their current locations in order to provide accurate location-based services. If the device contains a GPS chip, the device can determine its current location using GPS satellite data. But this process can take up to several minutes. Obviously, if the device does not have a GPS chip, no GPS location data will be available. To provide the high quality products and services that its customers demand, Apple must have access to comprehensive location-based information. To enable Apple mobile devices to respond quickly (or at all, in the case of non-GPS equipped devices or when GPS is not available, such as indoors or in basements) to a customer's request for current location information, Apple maintains a secure database containing information regarding known locations of cell towers and Wi-Fi access points – also referred to as Wi-Fi hotspots. As described in greater detail below, Apple collects from millions of Apple devices anonymous location information for cell towers and Wi-Fi hotspots.[3] From this anonymous information, Apple has been able, over time, to calculate the known locations of many millions of Wi-Fi hot spots and cell towers. Because the basis for this location information is the "crowd" of Apple devices, Apple refers to this as its "crowd-sourced" database.

The crowd-sourced database contains the following information:

> *Cell Tower Information:* Apple collects information about nearby cell towers, such as the location of the tower(s), Cell IDs, and data about the strength of the signal transmitted from the towers. A Cell ID refers to the unique number assigned by a cellular provider to a cell, a defined geographic area covered by a cell tower in a mobile network. Cell IDs do not provide any personal information about mobile phone users located in the cell. Location, Cell ID, and signal strength information is available to anyone with certain commercially available software.

Wi-Fi Access Point Information: Apple collects information about nearby Wi-Fi access points, such as the location of the access point(s), Media Access Control (MAC) addresses, and data about the strength and speed of the signal transmitted by the access point(s). A MAC address (a term that does not refer to Apple products) is a unique number assigned by a manufacturer to a network adapter or network interface card ("NIC"). MAC addresses do not provide any personal information about the owner of the network adapter or NIC. Anyone with a wireless network adapter or NIC can identify the MAC address of a Wi-Fi access point. Apple does not collect the user-assigned name of the Wi-Fi access point (known as the "SSID," or service set identifier) or data being transmitted over the Wi-Fi network (known as "payload data").

The crowd-sourced database does not reveal personal information about any customer. An Apple mobile device running Apple's mobile device operating system, iOS, can use the crowd-sourced database to (1) provide the customer with an approximate location while waiting for the more precise GPS location, (2) find GPS satellites much more quickly, significantly reducing the wait time for the GPS location, and (3) triangulate the device location when GPS is not available (such as indoors or in basements). The device performs all of these calculations in response to a request for location information from an application on the customer's device that has been explicitly approved by the user to obtain the current location, and the device requests from Apple the crowd-sourced database information needed for these calculations.[4]

The crowd-sourced database must be updated continuously to account for, among other things, the ever-changing physical landscape, more innovative uses of mobile technology, and the increasing number of Apple's customers. In collecting and maintaining its crowd-sourced database, Apple always has taken great care to protect its customers' privacy.

2. Downloading Crowd-Sourced Data to a Mobile Device

To further improve the speed with which the device can calculate location, Apple downloads a subset of the crowd-sourced database content to a local cache on the device. This content describes the known locations of Wi-Fi hotspots[5] and cell towers that the device can "see" and/or that are nearby, as well as nearby cell location area codes,[6] some of which may be more than one hundred miles away. The presence of the local cache on the device enables the device to calculate an initial approximate location before Apple's servers can respond to a request for information from the crowd-sourced database.

The local cache does not include a log of each time the device was near a particular hotspot or cell tower, and the local cache has never included such a log. For each Wi-Fi hotspot and cell tower, the local cache stores only that hotspot's/cell tower's most recent location information, downloaded from Apple's constantly updated crowd-sourced database. After a customer installs the free iOS software update, iOS will purge records that are older than seven days , and the cache will be deleted entirely when Location Services is turned off. The local cache is protected with iOS security features, but it is not encrypted. Beginning with the next major release of iOS, the operating system will encrypt any local cache of the hotspot and cell tower location information.

Apple issued a free iOS software update on May 4, 2011. Prior to the update, iTunes backed up the local cache (stored in consolidated.db) as part of the normal device backup if there was a syncing relationship between the device and a computer. The iTunes backup, including consolidated.db, may or may not have been encrypted, depending on the customer's settings in iTunes. After the software update, iTunes does not back up the local cache (now stored in cache.db). When a customer runs certain applications, those applications request location information from iOS. Because of a bug that existed prior to the update, even when Location Services was off, the device would anonymously send the IDs of visible Wi-Fi hotspots and cell towers, without any GPS information, to Apple's servers, Apple's servers would send back the known, crowd-sourced location information for those hotspots and cell towers (and nearby hotspots and cell towers), and the device would cache that information in the consolidated.db file. None of this downloaded crowd-sourced location information – or any other location information – was provided to or disclosed to the application.

The iOS software update fixed the bug that caused crowd-sourced location information to be downloaded to the device while Location Services was off. iOS will now delete any existing local cache from consolidated.db and, if Location Services is off, (1) Apple will not download any crowd-sourced location information to the device, regardless of whether a specific application requests that information, and (2) iOS will delete any cache of this information stored in cache.db.

3. Collections and Transmissions from Apple Mobile Devices

Apple collects anonymous location information about Wi-Fi hotspots and cell towers from millions of devices to develop and refine Apple's database of crowd-sourced location information. The mobile devices intermittently collect information about Wi-Fi hotspots and cell towers they can "see" and tag that information with the device's current GPS coordinates, i.e. the devices "geo-tag" hotspots and towers.

This collected Wi-Fi hotspot and cell tower information is temporarily saved in a separate table in the local cache; thereafter, that data is extracted from the database, encrypted, and transmitted – anonymously – to Apple over a Wi-Fi connection every twelve hours (or later if the device does not have Wi-Fi access at that time). Apple's servers use this information to recalculate and update the known locations of Wi-Fi hotspots and cell towers stored in its crowd-sourced database. Apple cannot identify the source of this information, and Apple collects and uses this information only to develop and improve the Wi-Fi hotspot and cell tower location information in Apple's crowd-sourced database. After the device attempts to upload this information to Apple, even if the attempt fails, the information is deleted from the local cache database on the device. In versions of iOS 4.1 or later, moreover, the device will not attempt to collect or upload this anonymous information to Apple unless Location Services is on and the customer has explicitly consented to at least one application's request to use location information.

4. Additional Location Information Collections

If Location Services is on, Apple collects location information from mobile devices under the following four additional circumstances.

First, as mentioned in Apple's April 27 response, Apple is collecting anonymous traffic data to build a crowd-sourced automobile traffic database with the goal of providing iPhone users an improved traffic service in the next couple of years. This information is temporarily stored in the local cache on the device, anonymously uploaded to Apple, and then deleted from the device. Second, Apple collects anonymous diagnostic information from randomly-selected devices to evaluate and improve the performance of its mobile hardware and operating system. For example, Apple may collect information about a dropped cell phone call, including the calculated location of the device when a call was dropped, to help identify and address any cell connection issues. Before any diagnostic information is collected, the customer must provide express consent to Apple. Apple cannot associate this information with a particular customer.

Third, Apple obtains information about the device's location (the latitude/longitude coordinates) when an ad request is made. The device securely transmits this information to the Apple iAd servers, the iAd servers immediately convert the latitude/longitude coordinates to a five-digit zip code, and the iAd servers then discard the coordinates. Apple does not record or store the latitude/longitude coordinates – Apple stores only the zip code. Apple then uses the zip code to select a relevant ad for the customer.

Finally, if a customer has consented to an application's collection and/or use of location information, iOS will provide current location information in response to a request from that application. iOS will provide that customer-approved application with the location of the device only; iOS does not provide applications with direct access to the local cache.

III. THIRD-PARTY APPLICATIONS AND THE IAD NETWORK

A. Third Party Applications

In July 2008, Apple launched the App Store where customers may shop for and acquire applications offered by third-party developers for the iPhone, iPad and iPod touch. Currently the App Store includes more than 350,000 third-party applications covering a wide variety of areas including news, games, music, travel, health, fitness, education, business, sports, navigation and social networking. Each application includes a description prepared by the developer regarding, among other things, what the application does, when it was posted, and, if applicable, what information the application may collect from the customer. Any customer with an iTunes account may purchase and download applications from the App Store. Developers do not receive any personal information about customers from Apple when applications are purchased. Only Apple has access to that information.

Third-party application developers must register as an "Apple Developer" by paying a fee and signing the iPhone Developer Agreement (the "IDA") and the Program License Agreement (the "PLA"). Registered Apple Developers gain access to the software development kit ("SDK") and other technical resources necessary to develop applications for mobile devices. The current PLA contains several provisions governing the collection and use of location-based information, including the following:

- Developers may collect, use, or disclose to a third party location-based information only with the customer's prior consent and to provide a service or function that is directly relevant to the use of the application;

- Developers must provide information to their customers regarding the use and disclosure of location-based information (e.g., a description on the App Store or adding a link to the applicable privacy policy);
- Developers must take appropriate steps to protect customers' location-based information from unauthorized use or access;
- Developers must comply with applicable privacy and data collection laws and regulations regarding the use or transmission of location-based information;
- Applications must notify and obtain consent from each customer before location data is collected, transmitted, or otherwise used by developers;
- If the customer denies or withdraws consent, applications may not collect, transmit, process or utilize the customer's location data; and
- Applications must not disable, override, or otherwise interfere with Apple-implemented alerts, including those intended to notify the customer that location-based information is being collected, transmitted, maintained, processed, or used, or intended to obtain consent for such use.

Developers that do not agree to these provisions may not offer applications on the App Store. Apple has the right to terminate the PLA if a developer fails to comply with any of these provisions.

Apple reviews all applications before adding them to the App Store to ensure, for example, that they run properly and do not contain malicious code. Apple, however, does not monitor applications after they are listed in the App Store, unless issues or problems arise.

B. The iAd Network

On July 1, 2010, Apple launched the iAd mobile advertising network. The network can serve ads to iPhone, iPod touch, and iPad devices running iOS 4, and the network offers a dynamic way to incorporate and access advertising within applications. Customers can receive advertising that relates to their interests ("interest-based advertising") and/or their location ("location-based advertising"). For example, a customer who purchased an action movie on iTunes may receive advertising regarding a new action movie being released in the theaters or on DVD. A customer searching for nearby restaurants may receive advertising for stores in the area.

As specified in the Policy and the relevant device SLAs, customers may opt out of interest-based advertising by visiting the following site from their mobile device: https://oo.apple.com. Customers also may opt out of location-based advertising by toggling the device's location-based service capabilities to "Off."For customers who do not toggle location-based service capabilities to "Off," Apple collects information about the device's location (latitude/longitude coordinates) when an ad request is made. This information is transmitted securely to the Apple iAd server via a cellular network connection or Wi-Fi Internet connection. The latitude/longitude coordinates are converted immediately by the server to a five-digit zip code. Apple does not record or store the latitude/longitude coordinates—Apple stores only the zip code. Apple then uses the zip code to select a relevant ad for the customer.

Apple does not share any interest-based or location-based information about individual customers, including the zip code calculated by the iAd server, with advertisers. Apple retains a record of each ad sent to a particular device in a separate iAd database, accessible only by Apple, to ensure that customers do not receive overly repetitive and/or duplicative ads and for administrative purposes. In some cases, an advertiser may want to provide more specific information based on a device's actual location. For example, a retailer may want its ad to include the approximate distance to nearby stores. A dialog box will appear stating: "'Advertiser' would like to use your current location." The customer is presented with two options: "Don't Allow" or "OK." If a customer clicks "Don't Allow," no additional location information is transmitted. If the customer clicks "OK," Apple uses the latitude/longitude coordinates to provide the ad application with more specific location information—the information is not provided to the advertiser.

In closing, let me again affirm that Apple is strongly committed to protecting our customers' privacy.

We give our customers clear notice of our privacy policies, and our mobile products enable our customers to exercise control over their personal information in a simple and elegant way. We share the Committee's concerns about the collection and potential misuse of all customer data, particularly personal information, and we appreciate this opportunity to explain our policies and procedures.

I will be happy to answer any questions you may have.

Exhibit A. April 27, 2011 Apple Q&A on Location Data

Mac
iPod
iPhone
iPad
iTunes
Support

April 27, 2011

Apple Q&A on Location Data

Apple would like to respond to the questions we have recently received about the gathering and use of location information by our devices.

1. Why is Apple tracking the location of my iPhone?
Apple is not tracking the location of your iPhone. Apple has never done so and has no plans to ever do so.

2. Then why is everyone so concerned about this?
Providing mobile users with fast and accurate location information while preserving their security and privacy has raised some very complex technical issues which are hard to communicate in a soundbite. Users are confused, partly because the creators of this new technology (including Apple) have not provided enough education about these issues to date.

3. Why is my iPhone logging my location?
The iPhone is not logging your location. Rather, it's maintaining a database of Wi-Fi hotspots and cell towers around your current location, some of which may be located more than one hundred miles away from your iPhone, to help your iPhone rapidly and accurately calculate its location when requested. Calculating a phone's location using just GPS satellite data can take up to several minutes. iPhone can reduce this time to just a few seconds by using Wi-Fi hotspot and cell tower data to quickly find GPS satellites, and even triangulate its location using just Wi-Fi hotspot and cell tower data when GPS is not available (such as indoors or in basements). These calculations are performed live on the iPhone using a crowd-sourced database of Wi-Fi hotspot and cell tower data that is generated by tens of millions of iPhones sending the geo-tagged locations of nearby Wi-Fi hotspots and cell towers in an anonymous and encrypted form to Apple.

4. Is this crowd-sourced database stored on the iPhone?
The entire crowd-sourced database is too big to store on an iPhone, so we download an appropriate subset (cache) onto each iPhone. This cache is protected but not encrypted, and is backed up in iTunes whenever you back up your iPhone. The backup is encrypted or not, depending on the user settings in iTunes. The location data that researchers are seeing on the iPhone is not the past or present location of the iPhone, but rather the locations of Wi-Fi hotspots and cell towers surrounding the iPhone's location, which can be more than one hundred miles away from the iPhone. We plan to cease backing up this cache in a software update coming soon (see Software Update section below).

5. Can Apple locate me based on my geo-tagged Wi-Fi hotspot and cell tower data?
No. This data is sent to Apple in an anonymous and encrypted form. Apple cannot identify the source of this data.

6. People have identified up to a year's worth of location data being stored on the iPhone. Why does my iPhone need so much data in order to assist it in finding my location today?
This data is not the iPhone's location data—it is a subset (cache) of the crowd-sourced Wi-Fi hotspot and cell tower database which is downloaded from Apple into the iPhone to assist the iPhone in rapidly and accurately calculating location. The reason the iPhone stores so much data is a bug we uncovered and plan to fix shortly (see Software Update section below). We don't think the iPhone needs to store more than seven days of this data.

7. When I turn off Location Services, why does my iPhone sometimes continue updating its Wi-Fi and cell tower data from Apple's crowd-sourced database?
It shouldn't. This is a bug, which we plan to fix shortly (see Software Update section below).

8. What other location data is Apple collecting from the iPhone besides crowd-sourced Wi-Fi hotspot and cell tower data?
Apple is now collecting anonymous traffic data to build a crowd-sourced traffic database with the goal of providing iPhone users an improved traffic service in the next couple of years.

9. Does Apple currently provide any data collected from iPhones to third parties?
We provide anonymous crash logs from users that have opted in to third-party developers to help them debug their apps. Our iAds advertising system can use location as a factor in targeting ads. Location is not shared with any third party or ad unless the user explicitly approves giving the current location to the current ad (for example, to request the ad locate the Target store nearest them).

10. Does Apple believe that personal information security and privacy are important?
Yes, we strongly do. For example, iPhone was the first to ask users to give their permission for each and every app that wanted to use location. Apple will continue to be one of the leaders in strengthening personal information security and privacy.

Software Update

Sometime in the next few weeks Apple will release a free IOS software update that:

- reduces the size of the crowd-sourced WI-FI hotspot and cell tower database cached on the iPhone,
- ceases backing up this cache, and
- deletes this cache entirely when Location Services is turned off.

In the next major IOS software release the cache will also be encrypted on the iPhone.

Press Contacts:
Natalie Harrison
Apple
harri@apple.com
(408) 862-0565

Natalie Kerris
Apple
nat@apple.com
(408) 974-6877

NOTE TO EDITORS: For additional information visit Apple's **PR website**, or call Apple's Media Helpline at (408) 974-2042.

End Notes

[1] The links take customers to http://www.apple.com/privacy, which customers may also access directly.

[2] Note that on March 31, 2011, Apple made two non-material updates to its June 21, 2010 Privacy Policy. Specifically, Apple updated: (1) the URL where users can login to their accounts to view and modify their preferences and contact information and (2) the mechanism provided to users to ask questions about the Policy.

[3] During this collection process, iOS does not transmit to Apple any data that is uniquely associated with the device or the customer.

[4] For devices running the iPhone OS versions 1.1.3 to 3.1, Apple relied on (and still relies on) databases maintained by Google and Skyhook Wireless ("Skyhook") to provide location-based services. Beginning with the iPhone OS version 3.2 released in April 2010, Apple relies on its own databases to provide location-based services and for diagnostic purposes.

[5] For each Wi-Fi hotspot, the location information includes that hotspot's MAC address, latitude/longitude coordinates, and associated horizontal accuracy number. For each cell tower, the location information includes the cell tower ID, latitude/longitude coordinates, and associated horizontal accuracy number.

[6] Cell base stations are grouped into "location areas" for network planning purposes, and each location area is assigned a unique "location area code." This "location area code" is broadcast by the cell base stations.

In: Mobile Devices ISBN: 978-1-62618-304-9
Editors: A. R. Walker, D. Brown Miller © 2013 Nova Science Publishers, Inc.

Chapter 7

STATEMENT OF JONATHAN ZUCK, PRESIDENT, THE ASSOCIATION FOR COMPETITIVE TECHNOLOGY. HEARING ON "PROTECTING MOBILE PRIVACY: YOUR SMARTPHONES, TABLETS, CELL PHONES AND YOUR PRIVACY"*

Chairman Franken, Ranking Member Coburn, and distinguished members of the Committee: My name is Jonathan Zuck, and I would like to thank you for holding this important hearing on privacy and the growing mobile devices marketplace. I am the president of the Association for Competitive Technology (ACT). ACT is an international advocacy and education organization for people who write software programs--referred to as application developers--and providers of information technology (IT) services. We represent over 3,000 small and mid-size IT firms throughout the world and advocate for public policies that help our members leverage their intellectual assets to raise capital, create jobs, and innovate.

Our community leaders are not political spokespersons—they are engineers; and I have drawn upon our membership's technical expertise and business concerns to inspire and inform these comments. Prior to this hearing, several Senators and their staff asked for information about the size, scope, and impact of this new apps ecosystem; my testimony here strives to answer

* This is an edited, reformatted and augmented version of a Presented May 10, 2011 before the Senate Judiciary Committee, Subcommittee on Privacy, Technology and the Law.

those questions as well as address concerns on privacy and security regarding mobile devices.

The new mobile apps world has sparked a renaissance in the software industry; small software companies are able to create innovative products and sell them directly to consumers. This is a radical departure from the era of up-front marketing costs, publisher delays, and piracy problems. The mobile app store has eliminated the longstanding barriers to entry that our industry battled for the past two decades. My goal today is to help explain how small business is building this exciting new industry, how what we are doing is helping consumers, and how the very real concerns about privacy must be dealt with holistically, rather than from a technology-specific perspective.

Finally, for this renaissance to continue, government action must be careful to preserve the opportunities for small businesses to innovate, experiment, and compete with dominant market players.

THE SMARTPHONE ECOSYSTEM IS CREATING JOBS AND OPPORTUNITIES IN TOUGH ECONOMY

The state of the U.S. economy is profoundly unsettled. Questions about job security, healthcare, and foreclosure have become dinner table conversation throughout this country. In the face of all of this turmoil, there has been a bright spot in economic growth: Sales of smartphones and tablets, such as the iPhone, the HTC Thunderbolt (running Google Android), the Samsung Focus (running Microsoft WP7) , the iPad, Xoom, and now RIM's Playbook, continue to outpace all predictions and are providing a huge growth market in a slumping economy. In fact, nearly one hundred million smartphones were shipped in the first quarter of 2011[1] marking a 79% increase in an already fast growing market.

Vendor	1Q11 Shipments	1Q11 Market Share	1Q10 Shipments	1Q10 Market Share	1Q11/1Q10 Change
Nokia	24.2	24.3%	21.5	38.8%	12.6%
Apple	18.7	18.7%	8.7	15.7%	114.4%
Research In Motion	13.9	14.0%	10.6	19.1%	31.1%
Samsung	10.8	10.8%	2.4	4.3%	350.0%
HTC	8.9	8.9%	2.7	4.9%	229.6%
Others	23.2	23.2%	9.5	17.1%	143.7%
Total	**99.6**	**100.0%**	**55.4**	**100.0%**	**79.7%**

Source: IDC Worldwide Quarterly Mobile Phone Tracker, May 5, 2011

Note: Vendor shipments are branded shipments and exclude OEM sales for all vendors.

Smartphones that run third party applications are creating opportunities for handset manufacturers like HTC, Apple, and Motorola, communications firms like Verizon and AT&T, and most especially for application developers like our members.

In 2008, Apple launched an —apps storel to provide a place for developers to sell independently developed applications for the iPhone. Since then, over 300,000 new applications have gone on sale with billions of applications sold or downloaded. The Android platform has recently exceeded the growth rate seen in the iPhone, totaling more than 200,000 applications with 10,000 new programs available each month. In 2010 we saw the release of Windows Phone 7 with its own applications store and an entirely unique user interface. Total unique apps across all platforms are expected to exceed 500,000 by the end of 2011.[3]

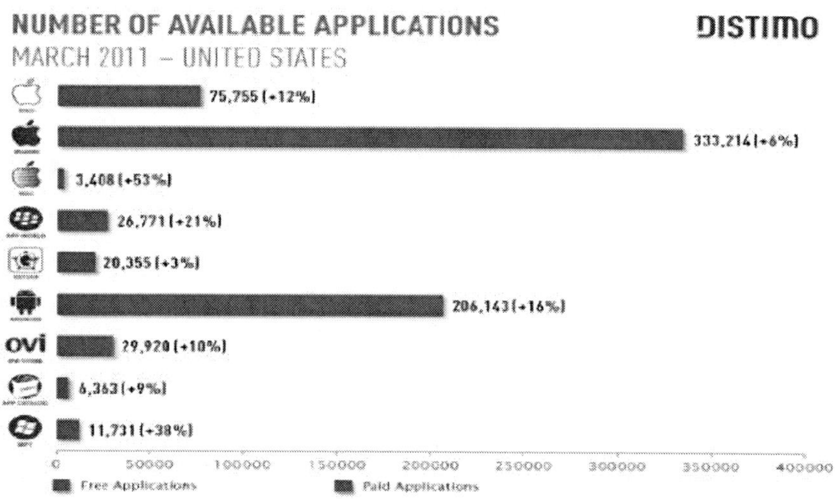

Possibly the most important thing we have noticed about the new apps world is how it has revolutionized the software development industry. It is nothing less than a rebirth. Startup costs of the modern app developer are a fraction of what they used to be just 10 years ago. Gone now are the costs of printing discs, manuals, marketing materials, contracts with retailers, onerous contracts with publishers, and contracts with credit card providers all once necessary to sell a single product. Distribution is now all digital. Those costs savings in distribution are now used to hire more developers and artists, thus creating more jobs across the country. With mobile and Xbox 360 apps, we have seen the return of the small "garage," independent developer focused on products that can be created and shipped in a matter of months. The apps store model creates a direct bridge between the customer and the developer. Our members tell us that being a developer has not been this exciting since the origins of the personal computer and software industry in the 70s and 80s.

So who is this new generation developer? What does an apps creator look like? To find out, ACT conducted surveys and focus groups within our membership and also analyzed the top 500 selling apps.

First, we learned mobile apps are overwhelmingly created by developers in small businesses. A review of the top 500 best selling applications show that over 85% are written by small businesses4; in a majority of cases, micro businesses with less than 10 employees.

Second, app developers are not just in California. During the dotcom boom of the 1990s, the majority of growth occurred in Silicon Valley while the rest of the country did not reap all of the benefits of the economic boom. Conversely, the recent growth of the mobile apps industry has led to job creation all across the United States. While California continues to have a large representation of apps developers, nearly 70% of the businesses are located outside of the state of California. The independent nature of this burgeoning industry allows developers to live almost anywhere, including Moorhead, MN, and Tulsa, OK[5].

Third, app development companies have low initial costs, but also have the ability to become a highly successful and sustainable business. ACT's members reported development costs ranging from $1,000 to upwards of $1,000,000. Given the wide range of our findings and those of other reports[6], it is better to view the cost of mobile apps in tiers. In tier one, a simple app with no real back end server based functionality can run in the low thousands; this category makes up a significant percentage of all the apps in various mobile stores. They may be single feature programs, vanity apps, or just irreverent apps like iBeer.

The second tier are the apps that provide multiple levels of functionality. Often working with data stored in a remote server to provide information/user generated content or advanced capabilities like writing and saving specialized documents, this tier runs from $30,000 to $100,000.

The final tier runs from $100,000 on up. This category is for apps that may need to tie into sophisticated inventory management systems, require specialized licenses for content, interface with business critical databases not just to read, but also write information, and finally, games with immersive environments where art and music costs can be significant.

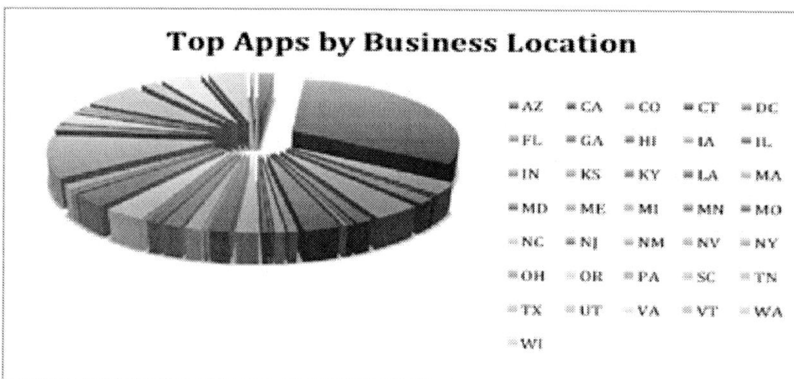

UNDERSTANDING THE REAL OPPORTUNITY FOR SMALL BUSINESS

To get a sense of the size of the market and potential opportunity, we must first understand the various business models underlying the mobile app market.

First, there are app developers who charge their customers to download their applications and/or charge them for purchases they make inside the app. For example, photography app Hipstamatic costs $1.99. If users want additional camera effects (Kodachrome or Holga, for instance) they can buy the add-ons in the application.

Second, some apps are supported either entirely or partly by advertising revenue. This is an increasingly important model especially as the Android platform grows in importance. Some applications charge for downloads and run advertisements inside the app itself.

Finally, many applications are given away free by larger companies in order to extend services to mobile devices or as marketing tools. From Citibank's online banking app to Pepsi's —Refresh Project‖ and Conde Nast's magazine apps, Fortune 1000 companies are increasingly offering mobile apps to their customers and potential customers. While large companies brand these apps, smaller companies with the expertise necessary to build world-class applications under tight deadlines usually build them.

Mobile App Stores

The exponential growth in app stores during the past few years is unprecedented. Apple launched the mobile app store arena with the iTunes App Store less than 4 years ago, soon followed by Nokia, Google, Microsoft, Amazon, and others. According to IHS, the worldwide market revenue of these app stores in 2010 was $2.15 billion, a 160% increase over 2009, and is expected to reach nearly $4 billion this year. Forrester Research estimates that the revenue created from customers buying and downloading apps to smartphones and tablets will reach $38 billion by 2015.

A growing percentage of revenues for app markets are coming from "in app purchases." According to Xyologic a company that indexes and analyzes app store data, 40 percent of game downloads are now free titles with in-app purchases. In March, it found there were more than 99.9 million downloads of free iPhone games from the App Store.

Yet revenues from app purchases and in-app purchases only represent a part of the overall opportunity for app developers. According to Xyologic, 80.8 percent of all app downloads in the month of March were free. While some of those apps relied on in-app purchasing for revenue, many others were supported by advertising or developed to support other brands and services.

Custom Mobile Development

The majority of the more than 600,000 free apps available across all app stores are not designed to be profitable on their own. They are designed as an extension to an existing service or a marketing program for an established or growing brand. Yet, the value of these apps and the jobs they create are completely missed by the revenue numbers of app stores and advertising platforms.

This translates into an tremendous number of job-creating opportunities for smaller app development shops. Forrester Research predicts this market to reach $17 billion by 2015.

Mobile Advertising Revenues

In-app mobile advertising is growing more slowly than revenues from app downloads and in-app purchases, but it is a particularly important revenue model for apps with enormous scale, or "eyeballs," like the hugely successful Angry Birds. In the games category, which represents around half the app market, the total revenue from in-app advertising was $87 million according to Juniper Research. Juniper expects that to grow to around $900 million by 2015.

It is also worth noting that the business model of the platform makes a difference in how developers pursue revenue. As shown in an earlier chart, the iOS store has more than 333,000 applications and nearly 70% of those are paid for up front. Google/Android, a company whose entire revenue stream and dominant market position is dependent on advertising, tends to push developers towards the advertising model, with only 30% of the 206,000 apps relying on direct payment to the developer.

The Future for Mobile App Developers

Even more important are the opportunities that lay farther ahead. Members of Congress all have BlackBerries and many have iPhones, Androids, or Windows Mobile devices as well. Yet, according to a recent Morgan Stanley report[7], most people haven't yet invested in such technology. True "smartphones" have around 25% penetration in the U.S.; in Asia, it may be as low as 6%. This represents a pathway for growth leading far into the future.

To understand just how important international sales are to the mobile apps market, one only needs to look at a comparison between the total number of users possessed by a combined AT&T / T-mobile (130 million wireless subscribers)[8] and China's number one wireless carrier, China mobile (584 million subscribers)[9]. Even if only 6% of China's mobile subscribers become smartphone users – and app purchasers – the market opportunity for U.S. software developers is huge.

How Location Based Information Helps Consumers

In the lead up to today's hearing, considerable attention has been directed at the type of information stored on smartphones. A misunderstood element in the public debate on this data collection is the essential role location information plays in the basic function of the device. People buy smartphones to have access to the Internet while they are mobile and a persistent connectivity is essential for this service.

When a smartphone tracks the location of its user, it is making a note to remind itself which access point or cell tower was used there to connect to the Internet. When a user returns to that area, the phone remembers this information. Each day most phone users travel the same route to work or to attend school and then return home to the same place. Keeping this data enables the smartphone to easily find an Internet connection providing efficient, constant online access. This is important for two reasons.

First is battery life. A phone uses a lot of power to search for a cell tower or wireless router. If it constantly needs to search for an Internet connection, it will deplete its battery many times more quickly than if it maintained a constant connection. Customers rate the importance of battery life very highly as a feature in the customer experience, so keeping a charge is a very important requirement of the phone. By maintaining a list of frequently visited locations, a smartphone avoids draining its battery in search of data connection points.

The other reason efficient connectivity matters is spectrum scarcity. The proliferation of smartphones has led to a crowded wireless spectrum, leading to potentially diminishing service quality. Wherever possible, wireless carriers are eager to connect users to wi-fi instead of their networks to provide faster connection speed and to lessen the burden on wireless networks. Carriers even provide their own wi-fi service for free to customers in densely populated areas to help alleviate the demand for wireless spectrum. By keeping track of the wi-fi and cell tower locations at frequently visited areas, the smartphone can allow users to automatically switch to wi-fi networks to provide constant, high quality Internet connectivity while diminishing the pressures on a crowded spectrum.

While location data is essential for phones to operate efficiently, consumers also love the smartphone services made possible using location-based technology. Many of the most successful apps or smartphone features have become popular based on knowing exactly where users are at any given time. And that's exactly how customers want it.

Anyone who has owned a smartphone has probably charted their location as a blue dot on their map app. Many also use those same programs to see where the traffic bottlenecks are before starting their evening commute. Some apps use location to help users find the nearest gas station, post office, parking garage, or coffee shop.

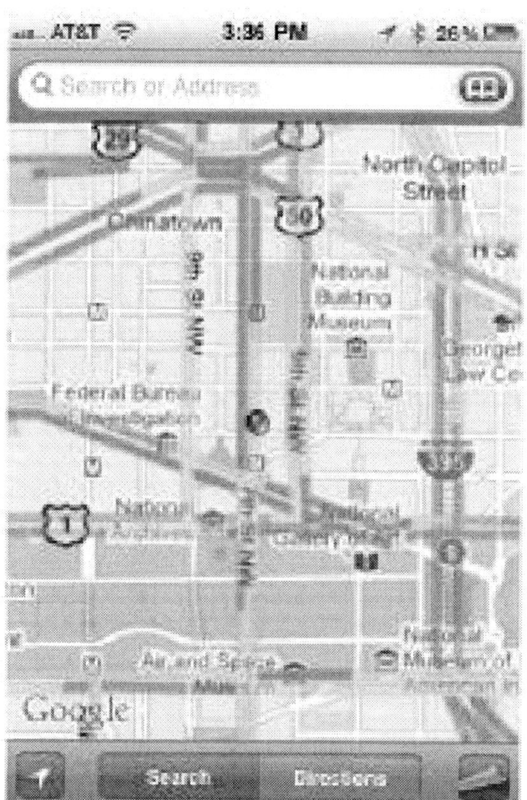

Map with Location and Traffic Data.

Open Table Reservations.

The OpenTable app adds location technology to its existing services to allow diners to find open tables at nearby restaurants, read reviews, and make reservations with a simple tap of the button. Using location information, the app can also provide step-by-step directions to the establishment.

Location services on smartphones have also changed the way we interact socially, creating a market for check-in features to tell your friends and family where you are. Facebook has an app with this feature and, within the last decade, has achieved a market valuation approaching $100 billion. Foursquare, an app which exclusively provides check-in services, has been valued at nearly half a billion dollars.

There is clearly big business opportunity in this marketplace. But location-based services and advertising offer a unique opportunity for Main Street businesses as well. Some apps, like RedLaser, allow users to scan the UPC code of a product and, using the smartphone's location data, find several local retailers nearby where it can be purchased.

Meanwhile, a user searching for a particular product or service on their smartphone can receive an ad from a local small business based on their current location data. These ads have the benefit of reaching potential customers at the exact time of a purchasing decision and cost far less than the newspaper circulars or the TV ads that big box stores are able to afford.

Similarly, local small businesses can also level the playing field with the national chain stores and Internet retailers through shopping apps like Groupon. This app serves 38 million North American subscribers who receive daily discounts at local establishments based on their location data.

While improving the core performance of smartphones, location data is also the building block for apps that users find useful and provide small businesses with opportunities to reach new customers. This data also contains information about the user which they may want to keep private so appropriate safeguards must be in place to ensure it is used in a manner with which consumers are comfortable.

Red Laser Bar Code Scan.

THE SMARTPHONE ID CONUNDRUM

Recent news stories have focused on the existence of unique identifiers attached to each smartphone. Known as a UDID number for iPhone and Android ID for Android based products, this is a number that serves as a unique token for each device. The *Wall Street Journal* article "What They Know - Mobile"[10] made special effort to note the transmission of this number by nearly every single application in the market. While highlighting the transmission of a "unique identifier" may make for good newsprint, the article unfortunately did not properly explain why developers transmit this number.

In order to help better explain the role this Smart Phone ID (SPID) number plays in the development and maintenance of mobile applications, ACT surveyed developers11 to find out how they currently used the SPID number. Respondents highlighted three key uses:

- Allows developers to control access to parts of the program without locking the user out completely (i.e., locking achievement levels in games, viewing paid subscriber content);
- Prevents piracy of applications, allows verification of ownership for updates to apps; and
- Allows management of access control for software testing and customer service.

Additionally, developers reported on several benefits to their customers specifically and consumers in general. Most often cited were:

- Working in concert with other stored data, the SPID makes it possible to have applications remember your favorites even when you buy a new phone;
- Helps content providers know when your device is on a wi-fi network instead of 3G - allowing them to send you HD or other high bitrate content; and
- Makes it easier to receive updates without annoying verification procedures.

At first glance, it would seem to make perfect sense to only allow the SPID to be shared with the app maker itself, but not with third parties. However, in today's world, many different companies work together to provide services to customers. For instance, when shipping a product via FedEx, the sender shares considerable personal information about the recipient with the (third party) shipper including contact information and purchased items. Similarly, small businesses rely on cloud computing to give customers a complete service offering in a cost-effective way. For game developers, a company like OpenFeint offers an easy way to keep track of scores and allows game users to interact with each other, saving app makers thousands of dollars in development time and ongoing infrastructure cost. This service needs to be able to tell devices apart.

Finally, developers felt that the usage restrictions and best practices for SPIDs were well documented, especially on Apple's iOS. As you can see from the documentation for the UIDevice.uniqueIdentifier[12], Apple gives plenty of advice to app makers on how to properly handle this information [emphasis added]:

> A device's unique identifier (sometimes abbreviated as UDID for Unique Device Identifier) is a hash value composed from various hardware identifiers such as the device serial number. It is guaranteed to be unique for each device. The UDID is independent of the device name. For devices that use a SIM (subscriber identity module) card, the UDID is independent of the SIM card.
>
> For user security and privacy, you must not publicly associate a device's unique identifier with a user account.
>
> You may use the UDID, in conjunction with an application-specific user ID, for identifying application-specific data on your server. For example, you could use a device-user combination ID to control access to registered products or when storing high scores for a game in a central server. However, if you are developing a game, you may want to instead use Game Center's player identifier key as explained in Game Kit Programming Guide.
>
> Important: Never store user information based solely on the UDID. Always use a combination of UDID and application-specific user ID. A combined ID ensures that if a user passes a device on to another user, the new user will not have access to the original user's data.

The key takeaway from this survey is that it is important, and often necessary, to keep devices separate and uniquely identified. Users may own many devices, multiple people may share devices (for example, family members), and others switch devices. Developers have different technical reasons to identify devices, but all come down to the same thing: enhancing the user experience. The developer's focus is in making the user's phone more convenient and useful.

While there may be some sinister ways in which the SPID can be illegally used, 99.9% of developers have the very best intentions. Specific instances of SPID abuse should be the focus of FTC action, not the very existence of such a valuable and valid tool.

Understanding the Existing Laws and Regulations

Regardless of how data protection is approached, it is critical to understand the protections available under existing federal and state laws and regulations. Consumer protection laws with technology-neutral legal standards can address data-privacy and data-security concerns regardless of whether they arise from undisclosed hacking, phishing, lost laptops, website data-collection, inadvertent peer-to-peer "sharing" of sensitive personal files, unauthorized wi-fi-snooping, recklessly designed social-networking applications like Google Buzz, art contests seemingly designed to enable the reverse-engineering of children's social-security numbers, or mobile apps.

Currently, the FTC Act gives the FTC broad authority to act against those who misuse data, regardless of the technology used. Specifically, Section 5 of the FTC Act directs the FTC to take action against any business engaging in "deceptive" or "unfair" trade practices.[13] The FTC's duty to halt deceptive trade practices authorizes the FTC to take law-enforcement action not only when a business violates explicit promises to consumers,[14] such as violations of stated privacy policies or terms of use, but also even when a business makes material omissions to consumers,[15] such as not telling consumers about the sharing of their collected information with third parties.

Similarly, the FTC's duty to halt unfair trade practices authorizes the FTC to take law-enforcement action when business practices cause injuries to consumers that are: substantial; not outweighed by countervailing benefits to consumers and competition; and could not have been reasonably avoided by consumers themselves.[16] For example, the FTC can take action against a business's failure to report a data breach.

Finally, it is critical to understand two points about consumer-protection laws. First, the FTC has real teeth if it finds that a company engaged in "unfair or deceptive practices," including assessing injunctive and civil penalties. Second, state consumer-protection acts grant state Attorneys General even broader substantive and remedial powers than those that federal law grants to the FTC. As a result, even were resource constraints or agency capture to preclude FTC action in a particular case, 50+ law-enforcement agencies would still have broad, technology-neutral authority to protect the privacy and security of consumers' data.

Consequently, the consumer-protection authority of the FTC and the State Attorneys General already authorizes and requires these law-enforcement agencies to patrol the Internet for companies that might violate their promises to consumers or cause them substantial harm. The FTC recently used such authority to protect consumer privacy by taking action against Google[17] and Chitika[18] for failing to properly handle consumers' information. Both companies now face twenty years of oversight and damage to their brands.

Existing consumer-protection laws thus already authorize both the FTC and state law enforcement agencies to police the entire range of products that connect to the Internet, including mobile devices, and to take action against the bad actors that ignore existing laws and will continue to ignore any future laws. This existing authority also ensures that good actors already have every incentive to behave reasonably and that bad actors have good reason to fear the existing legal consequences of their wrongdoing.

Given the existing authority of the FTC and the State Attorneys General, do we need additional regulation? ACT believes this is an open question, but one where consumer privacy protection should not be viewed through a limited, technology-specific lens. Instead, thoughtful, arduous, and considered discussion must take place on the role of personal data in the economy, the true interests of consumers, and the best interaction between citizens and the providers of products and services that use their data.

AVOIDING THE PATCHWORK PROBLEM; DEALING WITH DATA HOLISTICALLY

In periods of great technological change, both new opportunities and new challenges are created. More often than not, however, the seemingly new challenges are merely old issues illuminated under a new light.

Like the dot-com boom before it, the emergence of smartphones and mobile apps has renewed interest in the way corporations and governments collect and share data, most importantly, personal data. Yet, in both cases, these new technologies are simply bringing new light to issues surrounding personal data collection and use that have existed for decades.

There are genuine questions to be asked and considered with respect to the collection and use of personal data. How and when should people be told the data is being collected or when it is being shared? How should they be told? Should people be able to modify data that is collected about themselves? Should people be able to delete data about themselves or otherwise control how it is used? Asking these questions only in the context of smartphones and mobile apps ignores the larger picture. The technology used to collect the data is much less significant than the important questions about the process and behavior of those collecting it.

First, the data collected by apps developers is an almost infinitesimal piece of the global collection of personal data. From credit card companies, to warranty cards, to loyalty programs, companies have been collecting data on their customers long before the Internet or smartphones came around. Not only do other companies collect the same data as smartphone apps, but they have exponentially larger collections of personal data already at their disposal. Information brokers like Epsilon and Google collect, retain, and share far more information than all mobile apps combined.

Even the collection of location data that has been singled out in recent press reports is not unique to smartphones and mobile apps. Standalone commercial GPS providers like TomTom or GPS-based safety services like OnStar collect this information on their users. Your EZ Pass technology for wireless payment of highway tolls also collects and stores location data. More recently, Google has been collecting personal information while mapping home and business wireless networks. In nearly every instance, these companies may share that data with third parties.

All of this reminds us that isolating and regulating one specific technology is not the answer to the broader questions surrounding the collection and sharing of personal data. Given the enormity of existing data collections and the number of ways it is amassed, focusing exclusively on one technology – particularly the newest and least established – is a symbolic gesture that does not solve the underlying problem, but creates the false sense that the problem has been solved and the need for thoughtful debate and policy consideration is over. However, focusing instead on regulations of behavior and data usage, it then applies to everyone, regardless of means of collection and sharing.

Finally, perhaps the most dangerous problem is that when regulation focuses solely on new technology, it discriminates against small businesses. Whenever we are talking about new, disruptive technologies, we are most often talking about small businesses. Revenue models, customer expectations, and efficiency opportunities are all still emerging, and it is small businesses that perform that service. Lots of businesses start, a very small number survive, but in the end, we know what works, and then the large businesses get involved. To stunt the growth of a new, experimental market is to discriminate against the very small businesses on which we rely to lead innovation and growth in the American economy.

CONCLUSION

The future of the digital marketplace looks bright for small business, so long as the marketplace remains dynamic and competitive. This is a more than $10 billion opportunity for small business across the United States. Barriers to entry in the marketplace are currently low, and our members are very excited about the future – according to ACT's Board President, Mike Sax, —Programming is fun again! While there are important questions that need to be discussed on personal data collection, retention, and sharing, limiting this question solely to smartphones and mobile apps would be ineffectual and counterproductive. The use of location information and smartphone ID's are providing immense value to consumers. Whether it's the ability to make dinner reservations or find directions to the nearest hardware store, our members put a value on creating a product that improves the lives of their customers.

Banning the collection of location data would essentially outlaw these beloved consumer apps while doing nothing to address the big questions about data collection and how that data is used. That is why ACT believes that Congress must take a holistic approach to privacy that does not single out any one technology, especially nascent ones. We need to outlaw bad behavior, not good technology. I hope that the committee will continue to focus the spotlight on the contribution small business makes to the future of the digital economy and the way government can do a better job to encourage that productive future. Thank you for your time and consideration on this important topic.

End Notes

[1] Mark Kurlyandchik, IDC: Nokia Remains Top Smartphone Vendor Worldwide, DailyTech, May 6, 2011.

[2] Id

[3] http://d2omthbq56rzfx.cloudfront.net/wp-content/uploads/2011/04/Distimo-survey-201103-app-stores-count.png

[4] ACT analysis of top 500 selling apps, some discrepancies exist due to lack of verifiable employment data and apps created by a developer who has significant investment from a larger company. Some apps branded for a larger company are in fact developed by small firms subcontracted to build the application. Sample size of 408 applications, from "top apps" on March 25, 2011.

[5] ACT study of top selling apps as of March 25, 2011. ACT members Chalk LLC in Moorhead, MN, and Permafrost Software in Tulsa, OK.

[6] http://appmuse.com/appmusing/how-much-does-it-cost-to-develop-a-mobile-app/

[7]http://www.morganstanley.com/institutional/techresearch/pdfs/2SETUP_12142009_RI.pdf

[8] http://www.siouxcityjournal.com/business/local/article_f24b5818-ea11-5f04-b0b0-d7 bbd020 55b0.html

[9] http://www.wirelessweek.com/News/2011/01/Carriers-Subs-Reach-842M-China-Mobile/

[10] http://blogs.wsj.com/wtk-mobile/

[11] ACT April 28 questionnaire to members working on at least one mobile platform. Question: How do you currently use UDID/Android ID in your development process?

[12] http://developer.apple.com/library/ios/#documentation/uikit/reference/UIDevice_Class/ Reference/UIDevice.html

[13] 15 U.S.C. § 45

[14] Id.

[15] FTC, Policy Statement on Deception (Oct. 14, 1983) available at http://www.ftc.gov/ bcp/policystmt/addecept.htm.

[16] 15 U.S.C. §45(n); see also FTC, Policy Statement on Unfairness (Dec. 17, 1980) available at http://www.ftc.gov/bcp/policystmt/ad-unfair.htm.

[17] In the Matter of Google Inc., a corporation, FTC File No. 102 3136.

[18] In the Matter of Chitika, Inc., a corporation, FTC File No. 1023087.

INDEX

J

L

M

N

S

T

U

V

W

X

Y